D1084209

Florida Critters

F L O R I D A
CRITTERS

Common Household and Garden Pests
of the Sunshine State

BILL ZAK

Taylor Publishing Company
Dallas, Texas

Copyright © 1986, Taylor Publishing Company
1550 West Mockingbird Lane, Dallas, Texas 75235

Library of Congress Cataloging-in-Publication Data

Zak, Bill.
    Florida critters.

    Includes index.
    1. Insect pests — Florida — Identification.
2. Garden pests — Florida — Identification. 3. Household
pests — Florida — Identification. 4. Insect pests —
Control — Florida. 5. Garden pests — Control — Florida.
6. Household pests — Control — Florida. I. Title.
SB934.5.F6Z35 1986   648′.7′09759   86-1951
ISBN 0-87833-531-5

Designed by Bonnie Baumann

Printed in the United States of America

First Edition.   9 8 7 6 5 4 3 2 1

———————————For Jean Elizabeth Zak———————————

My wife, critic, helper and avid supporter, who
again patiently tolerated a sometimes impossible
husband and shared him with a computer for
several months while this book was written.

____ Acknowledgments _____

The manuscript portion of this book was reviewed for technical validity by Dr. Wayne N. Dixon, Florida Department of Agriculture, Division of Forestry. He also provided the information about scientific classification. It is so nice when a working relationship develops into a fine friendship, and this was my experience with Dr. Dixon. I am forever grateful for his generosity, expertise and friendship.

I wish to offer most sincere thanks to the University of Florida, Institute of Food and Agricultural Sciences; the Florida Department of Agriculture, Division of Forestry; the Florida Department of Agriculture, Division of Plant Industry; Southern Forest Insect Work Conference, Asheville, North Carolina; the United States Department of Agriculture — Photography Division, Washington D.C.; Texas Forest Service, a part of the Texas A&M University System; Texas Agricultural Extension Service (Entomology Department) of the Texas A&M University System; Big State Pest Control, Houston, Texas; Van Waters & Rogers, division of Univar, San Mateo, California; Ortho Information Services, Chevron Chemical Company, San Francisco, California; Dow Chemical, U.S.A.; and Manatee County Mosquito Control District, Palmetto, Florida. These public agencies and private companies made available the photographs shown in this book. I wish, also, to express appreciation for the efforts of innumerable entomologists and other employees of these agencies and corporations who caused to be published myriad technical documents that provided other data sources.

Individuals to whom I shall always be grateful: William J. Spitz, Chairman of the Board, Big State Pest Control; Dr. Robert C. Bullock, University of Florida, Institute of Food and Agricultural Sciences; Dr. Joe Eger, Dow Chemical U.S.A.; Lawrence M. Rhodes, Director, Manatee County Anti-Mosquito District; Harold A. Denmark, Dr. Avas B. Hamon, Dr. Franklin W. Mead, Dr. G. Wallace Deckle, V. Jane Windsor, Jeffery Lotz, Dr. G.B. Edwards, and Dr. Lionel A. Stange, Florida Department of Agriculture, Division of Plant Industry; Dr. Donald E. Short and Dr. Philip G. Koehler, University of Florida, Institute of Food and Agricultural Sciences; and Maggie (Grandma) Zak and W.J. (Grandpa) Zak, my first teachers. My most sincere thanks to you all.

Bill Zak

____Code for Slide Sources_____

DOF	Wayne N. Dixon (PhD); Florida Department of Agriculture, Division of Forestry.
DPI	Harold A. Denmark, Avas B. Hamon (PhD), Franklin W. Mead (PhD), G. Wallace Deckle (PhD), V. Jane Windsor, Jeffery Lotz, G.B. Edwards (PhD), Lionel A. Stange (PhD); Florida Department of Agriculture, Division of Plant Industry.
SFI	Southern Forest Insect Work Conference.
IFAS	Donald E. Short (PhD), Philip G. Koehler (PhD), Robert C. Bullock (PhD); University of Florida, Institute of Food and Agricultural Sciences.
USDA	Clemson University Extension Service in cooperation with the Federal Extension Service USDA.
TAEX	Texas Agricultural Extension Service (Entomology Department) of the Texas A&M University System.
TFS	Texas Forest Service.
MCMCD	Lawrence M. Rhodes, Director, Manatee County Anti-Mosquito District.
BS	William J. Spitz, Chairman of the Board and CEO, Big State Pest Control.
DOW	Joe Eger (PhD); Dow Chemical, U.S.A.
VWR	Van Waters & Rogers, division of Univar.
ORTHO	Ortho Information Services, Chevron Chemical Company.
JEE	James E. Everett.

Contents

____Foreword_____

If you live in Florida the chances are high that you have more than a passing acquaintance with a few of the more common pests. The odds are also very high that you will encounter pests which don't enter into the everyday over-the-fence or cocktail-patio conversation. Florida simply has a bundle of well-known and lesser-known pests that are constantly nibbling or sucking on our trees, shrubs, bushes, flowers, vegetables, fruits, houses, groceries, and furniture! Some even bore through lead, concrete, and drywalls. Pests make no social, financial, color, creed, or religious distinctions. Everybody is a likely victim under the right environmental conditions. Simply put: Florida is blessed with nice people, balmy tropical weather, good food, and a plethora of pests.

Sounds ominous. Before you despair, there is help. The book you have in your hands will go a long way towards rectifying your pest problem. Bill Zak has put together a volume of information that will help you identify who the pest is — I like Bill's generic term of "CRITTERS" — a very important first step. Once you have put a moniker on the critter, you can read about its biology. Plus enjoy Bill's humor which may reduce some of your anxieties. Keeping in mind how the critter operates, you can then put together a control program (not always by pesticide) that will suppress or eliminate the critter problem. Not all pests require absolute eradication. Oftentimes, so much pesticide would be required that it would pose a severe environmental hazard and most likely would be illegal. In fact, to attempt to kill all pests may only give you a worse problem with another critter!

Taxonomic information is provided for the reader with an interest in the critters other than immediate destruction. Your local library will have several books that present additional information on the pests. Common names aren't always the same throughout a region, thus the more technical information is very helpful to ensure we all are talking about the same critter or group of critters.

I know you will find Bill's book a boon, and perhaps it will help you keep your sanity while you wonder what you did to deserve your critter's company.

Wayne N. Dixon, PhD, RPE

Introduction

The major problem in pest control is not the selection and use of a pesticide, but is rather the proper identification of the pest. Also keep in mind that some critters do noble deeds and should be spared. Which wear halos — which wear horns? This book is intended to help homeowners make those decisions. On the following pages, you will see suggested controls for the "bad guys." Be advised that these are intended to be only clues to controls. I don't suggest they are the only controls, or necessarily the best controls in any given situation. So, with that caution, let's proceed to meet our friends and foes. Welcome to the Critters, some common garden and household pests of Florida.

My grandpa, Adolph Stasny, was a farmer, country store owner and cattle raiser near the small community of Smetana, Brazos County, Texas. A necessary family food source back in those days was the vegetable garden. Families were large and so were the gardens. My mother, known to some of you as Grandma Zak, tells how garden insects were controlled when she was a little girl some seventy-odd years ago.

When it got "good-and-dark," Grandpa gave each of the eleven children a kerosene lantern and a tin can with a little "coal-oil" in the bottom. Everybody took a row and, by lantern light, hand-picked every insect in sight, depositing same in their coal-oil can. It took quite a while to do this because the garden was somewhat over an acre in size. After the garden was scoured for insects, everyone turned in their cans to Grandpa who then dumped the critters in a pile and burned them. This ritual was repeated as needed. And those were the good ol' days? I wonder how many of us would opt to garden if this technique were necessary today. Well, to coin a phrase, we've come a long way, baby!

Bill Zak

_____Pest Control_____

Less than one-half of one percent of all insects are considered pests of plants. Plant pests are controlled in a number of ways. Mother Nature provides factors that influence insect numbers without any effort by man; then we have biological control . . . natural control with an assist by man; then there is what might be called legal control, exclusion by quarantine, to prevent the introduction of critters; also there is control by cultural practice, by mechanical and physical practices; and, finally, by the use of chemicals.

NON-CHEMICAL CONTROLS

NATURAL CONTROLS are frequently not thought of. Not many insects live in all climates — arctic, temperate and tropical. Cold winter temperatures restrict the range of some insects; summer heat limits others. Some insects prefer a warm, moist climate; others like it warm, but dry. Some insects can fly, or can be carried by the wind over long distances. Geographic barriers, like lakes, rivers and mountain ranges, can check the spread of insects; the nature of the soil deters others, for example, wireworms flourish in poorly drained soil — nematodes in sandy soil. Birds, moles, skunks, snakes, lizards, newts, salamanders and toads eat many insects. Birds consume enough to more than compensate for the strawberries, tomatoes, grapes or blueberries they also eat. Bird netting over your vulnerable fruiting plants will usually tip the scale in your favor. Skunks and armadillos eat a lot of grub worms, but in the process root holes in turf and garden. Skunk and armadillo critters are nocturnal. They can be easily frightened and will leave of their own volition if given the opportunity, but be wary of a skunk if he raises his tail; the hunted then becomes the hunter. Most of us know not to mess with a skunk, but your dog might have to learn the hard way. (Note: Tomato juice will neutralize skunk odor, but don't try to feed it to your dog if confrontation results in his getting skunk-sprayed; he won't drink it. Bathe him in it.)

BIOLOGICAL CONTROLS consist of artificial restoration of the balance of nature; Bacillus thuringiensis as a control for certain caterpillars is a good example. Bringing in insect predators and parasites, encouraging birds, introducing certain nematodes or fungi to prey on the undesirables will be effective to varying degrees.

CONTROL BY EXCLUSION can be either legal or voluntary. The Plant Quarantine Act of 1912 provides that whenever it is deemed necessary, in order to prevent the introduction of any dangerous insect or plant disease, the federal government will have the power, after a public hearing, to prohibit the importation, or shipment interstate, of any class of plants or plant products from any country or locality and from any state or territory in this country. Such specific prohibitions are called quarantines. It was never expected that a quarantine could keep out a certain pest forever, but the expense of the inspection service is justified if the insect is excluded for a period of time sufficient for us to learn its life history, to develop controls and to introduce its predators or parasites.

CULTURAL CONTROLS may simply be a matter of good housekeeping in the garden. Sometimes it does not even require labor, but rather a little planning based on the knowledge of life histories of certain insects. Most insects attack only a small number of related plants. Cabbage worms chew members of the cabbage family; the squash vine borer prefers the cucurbit group; Mexican bean beetles don't like much except beans; sweet potato weevils have an ongoing love affair with sweet potatoes. By switching locations of crops (crop rotation) you can starve out certain pests, or at least keep them from building up huge populations.

Commercial farmers have taught us trap-cropping (planting an early, expendable crop to attract an insect pest for elimination). Soil cultivation destroys some critters; tilling the vegetable patch in the winter exposes some larvae and pupae and they are killed by the cold; tilling in the summer kills some by exposure to hot temperatures. Timing of planting is another good cultural control for insects. Green beans planted early often mature between broods of the Mexican bean beetle; early summer squash may come along ahead of the squash vine borer; late corn is less apt to be injured by the European corn borer.

Often the character of the foliage keeps off pests. Leafhoppers usually prefer varieties with smooth leaves, but the azalea whitefly restricts operations to varieties that have hairy leaves. The effectiveness of these control-tricks will depend upon your knowledge of and experience with your individual garden . . . further proof that good-gardening is really an art and not a science.

Good housekeeping in the garden is tremendously important in controlling pests. Clean up all plant parts after harvest, compost what you can and burn anything capable of causing trouble later, for instance, the limbs severed by the twig girdler. Get rid of weeds and plant debris to avoid offering a haven to cucumber beetles, Mexican bean beetles and Colorado potato beetles as well as a multitude of other pests. I like to shred all plant matter larger than pine needles before composting.

CONTROL BY MECHANICAL AND PHYSICAL MEASURES consists of barriers being erected between the plant and the pest. A wire fence will keep out rabbits and other animals; bulbs planted in wire baskets will protect against mice and moles; wire shields will keep dogs off shrubbery; hardware cloth protects orchard trees from rats, mice and squirrels; metal bands placed at the base of trees will fend off beavers; nylon bird netting properly placed will confound the brightest mockingbird and spare your ripened strawberries, tomatoes and fruits; a cardboard cylinder or tin can will shield young transplants from cutworms. Dare to be ingenious. Remember, all is fair in love, war and gardening.

CHEMICAL CONTROLS

We have more insecticides, fungicides, herbicides and technology available to us than ever before in our history. We also have the greatest number of safety regulations controlling their use. And we also enjoy the most abundant and varied food supply of any known civilization to date. So, maybe everything within the chemical pest control industry is not all *right*, but certainly everything can't be all *wrong*.

The Federal Environmental Pesticide Control Act of 1972 in part prohibits the application of any pesticide in a manner inconsistent with its labeling. This means that a pesticide cannot be recommended unless it is registered for the specific pest or target site. Such registration is accorded only after intensive scientific investigation is responsibly performed and documented, attesting to the relative safety of said pesticide. In other words, it's a case of being guilty until proven innocent. So, READ YOUR LABEL; don't arbitrarily increase suggested dosages; if your particular pest problem or target site is not described on the label, ask a responsible authority. Don't guess.

Pesticide label clearances are subject to change. In the event you encounter a mandate that is inconsistent with any application recommendation in this book, make a note on the Notes page and follow the new rules. Remember, the pesticide user, YOU, is always responsible for the effects of pesticides on your own plants or household goods as well as for problems caused by drift from your property to other property or plants. Don't be careless.

PETROLEUM OILS (Dormant, Summer and Superior Oils) are insecticidal in that they suffocate the egg or the immature form or the adult form of pests. They are not toxic to the nervous system of animals; therefore, they are safe for use by man. They are contact insecticides that interfere physically, rather than chemically, with respiration. If the user will read the label and follow directions, few problems will be incurred.

DORMANT OILS are the heaviest and are formulated for use on plants that are dormant. These oils should be applied as late in the dormant season as possible, but before plants enter the bud-break stage in the spring. DO NOT USE A DORMANT OIL DURING THE GROWING SEASON, unless the label expressly states that it can be used.

SUMMER OILS are lighter than dormant oils and are formulated for use during spring and summer. Follow label directions closely. Don't cheat or you will be sorry.

SUPERIOR OILS are the most stringently refined of all and are excellent for horticultural use. They are so named because they are refined under certain specifications which allows them to be used in any season provided weather conditions permit their use. Superior oils are refined primarily for use on trees during the growing season, but they may also be used as a dormant oil.

Certain plants are highly sensitive to oils and should not be treated; these include hickory, certain conifers, ferns, palms and African violets. Many cacti and succulents are also susceptible to injury. Read your label. Most labels recommend that oils be used at temperatures between 40 degrees and 90 degrees F. This is the preferred temperature range. Petroleum oils are probably the most effective of all the pesticides against scale insects. When using oils, spray thoroughly to get coverage of the entire plant; if this is achieved no further spraying should be necessary for several months or a year.

THE CRITTERS

Ants

FLORIDA CARPENTER ANTS

Florida Carpenter Ants, also known as Bull Ants or Bulldog Ants, are big critters . . . ¼ to ½ inch long. Compared to the other out-of-town carpenter ants which are usually all reddish-brown or black in color, this Florida model has a reddish-yellow head and thorax with a black abdomen. This guy has class.

Although these giant ants prefer to nest in logs, stumps or dead trees where the wood contacts the soil and moisture is plentiful, they will also infest structural wood of buildings. They especially prefer construction wood that has been damaged by termites or by decay, but once established may very well invade sound wood. Carpenter ants do not eat wood, but they remove it and eject it as sawdust in the process of constructing galleries for nests. These galleries are kept very smooth and clean . . . an almost sandpaper appearance. Other wood-infesting insects do not keep their galleries this clean.

Carpenter ants feed on honeydew from sapsucking insects, on dead and live insects and on food wastes in and around houses. A typical nest will contain several thousand individuals and can exist for several years if not treated. Deadwood areas of living trees are occasionally infested.

Florida Carpenter Ant = Order: Hymenoptera; Family. Formicidae

CONTROL CLUE

Spray baseboards, windowsills, door frames and other surfaces where carpenter ants crawl with Dursban, Diazinon or Baygon. Find the nest and treat it. Often children like to watch ants, so they may be useful in tracing the ant trail back to the nest. This is a good way for kids to make a "buck" and for you to get kids out of your hair for an hour or so. Everybody wins . . . except for the ants, but that's the idea.

Florida Carpenter Ants
(DPI)

Carpenter Ant damage
(Ortho)

Florida Carpenter Ant galleries
(DOF)

Carpenter Ants
(Ortho)

FIRE ANTS

Fire Ants are hell . . . to live with or to control. Forgive the crudeness, but I don't know a better way to say it. Five species occur in the United States. Two species are of particular concern in Florida: Tropical or Native Fire Ants and Red Imported Fire Ants, the latter being of most concern. The total area infested in Florida is estimated at over 30 million acres and these ants are found in every county. Imported fire ants are ⅛ to ¼ inch long and are reddish-brown to black in color. They are social insects and live in colonies which may have up to 250,000 members. They are capable of making mounds that can be 8 inches to 3 feet high and up to 4 feet wide. Since it is neither necessary nor desirable to treat for native fire ants, it is important to make an identification before attempting treatment.

Fire ant colonies are composed of a queen, winged males and females (virgin queens), workers and brood (the little ones, including eggs). The winged reproductives leave the colony on mating flights mainly in late spring and summer, although swarming may occur at anytime of the year. The ants mate during flight, then females land to begin new colonies. Most females fly or are blown by wind less than one mile from the nest, but some may travel 12 miles or more. They prefer to nest in open, sunny areas, usually where you prefer to have your picnic or ball game or frisbee match.

Imported fire ants spread naturally through queen movement. This circumstance may be accomplished by crawling, by rafting downstream on or in logs, or by travelling aboard cars, trucks or trains. Shipments of soil or nursery stock from an infested area may relocate an entire ant colony or nest.

Agricultural losses from this pest are significant, but in urban areas imported fire ants are a formidable problem also. They invade lawns, parks, playgrounds, schoolyards, cemeteries, golf courses and homes. That fluffy pile of dirt in your water meter box? Fire ants? These villains attack anything that disturbs their mound or life style. Symptoms of the fire ant sting include burning and itching, often followed by the occurrence of a white sore or pustule that may leave a permanent scar after healing. *Venom of this ant is unlike that of any other stinging insect.* Hypersensitive persons suffer chest pains or nausea and may even lapse into a coma from just one sting. Persons who react severely to a fire ant sting should see a physician immediately.

Fire Ants = Order: Hymenoptera; Family: Formicidae

CONTROL CLUE

The technique used in applying an approved fire ant insecticide will be as important as the selection of the insecticide. To control any of the species effectively, an understanding of their habits, especially their mound-building activities, is critical. Fire ant mounds will be more apparent in the spring, summer and early fall following rainy periods. These ants do not prefer extremely wet soil; therefore, they increase their mound size and move their brood up into the drier soil of the mound. This habit makes a colony more vulnerable to insecticide control measures.

The most effective ant control is a bait application. Worker ants will gather the poisoned food and feed it to the queen. Killing the queen is the key to colony elimination. Drenches do not always kill the queen, but if you drench properly, you will have increased your chances. The most effective drench procedure is to dilute an approved insecticide in water. Apply this solution (usually a gallon or so depending upon mound size) to simulate a gentle rain. A sprinkler can will be especially useful for this procedure. Thoroughly wet the mound and surrounding area to a diameter of 3 to 4 feet.

It is important not to disturb the mound before or during treatment. Mound disturbance causes part of the colony to move away from the chemical. If most of the colony is in the mound and above ground level, treatment should be successful. Surviving workers may remain for a week or two, but will ultimately die. If mounds are missed during the first application, re-treat when ants are noticed. This procedure is usually successful on cooler spring or fall days and during winter months. During very dry periods, most members of a fire ant colony are well below ground level. Applying an insecticide at this time does little more than move the colony entrance to another location. To overcome the dry condition, water-soak the soil around the mound. This procedure drives the colony up into the drier soil. Now, proceed with control procedures previously described.

Several products are labeled for fire ant spot treatment and will likely contain Dursban or Diazinon. They are available in liquids and granules. Use according to label directions. Want to use a bait? How about Amdro. Read the label.

Red Imported Fire Ant mound
(Dow)

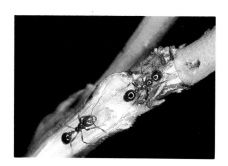

Red Imported Fire Ants on citrus
(Dow)

Red Imported Fire Ant damage and
tumulus
(DOW)

Red Imported Fire Ants gathering food
(VWR)

Imported Fire Ant larva, pupa, worker,
drone and queen
(VWR)

FLORIDA HARVESTER ANT

Here's the ant that you will painfully, but nostalgically, remember if you grew up as a barefooted country kid — the "Red Ant" that would hunch-up on your big toe and inflict such a painful sting that it ruined your Sunday afternoon over at Grandma's house. This is the same critter that would make your dog yelp with pain if one happened to curl-up in a stinging knot between his toes. One red-ant sting would forever be a vivid reminder to respect that big, bare, sandy mound with the hole in the top. This is the home of the Florida Harvester Ant . . . where you dropped the dead snake, whose remains would mysteriously disappear after a couple of days.

Harvester Ants live in large, deep colonies with the queen being ensconced 2 to 6 feet underground. These ants are also known as Agricultural Ants because of their preference for seeds and other plant materials which foraging worker ants gather and haul into the nest with breakneck speed. All vegetation around their mound will be cleared, leaving big bare circles that may be 2 to 12 feet in diameter. Florida Harvester Ants are ¼ inch to ½ inch long and are red to dark brown in color.

Florida Harvester Ant = Order: Hymenoptera; Family: Formicidae

```
_____ CONTROL CLUE _____

These are "country" ants and they seldom require control, but
if their presence poses a danger to animals or humans, or if
their mound proves to be undesirable, pour a liberal amount of
Dursban liquid down the hole in the center of the mound.
```

A close relative of the Florida Harvester Ant (Good enough for your purpose, unless you are a harvester ant!)
(VWR)

Red Harvester Ant mound
(VWR)

HOUSEHOLD ANTS

Many different ants are commonly found in Florida and they readily invade homes for feeding or nesting or both. Homeowners become alarmed when ants are found infesting food or other items and ordinarily respond by drenching the area with enough insecticide to kill a mule. Now, it's O.K. to "sic 'em,"but don't lose your head. You have a right to be upset: even the names of common Florida house and yard infesting ants are enough to make chills. Listen to what you're up against: Crazy Ant, Pharaoh Ant, Thief Ant, Little Fire Ant, Little Black Ant, Pyramid Ant, Fire Ants, Carpenter Ants, Harvester Ants and others. This lineup frightens me also, but applying extensive, undirected insecticide treatments in the house is unsatisfactory because this action kills only a small number of the invading worker ants and usually does little to affect the colony which is the source of the workers . . . and your ant problem.

Ants come in your home because food, water and shelter are available there. They are one of the most successful insect groups because they so readily adapt to changing environments. Most ants prefer to nest in soil or outdoor wood, but your home offers many desirable locations for nesting; cracks and holes in brick veneer, wall voids and structural wood are sites commonly sought out and exploited. At certain times of the year, mature ant colonies produce winged reproductives which swarm from the nest in great numbers. When these swarmers emerge within or near a building, they may be confused with termite swarms, but if they are ants, the nest is usually located in proximity to the swarm. Stay cool.

Household Ants = Order: Hymenoptera; Family: Formicidae

_____ **CONTROL CLUE** _____

The most effective ant control lies in finding the nest and treating it with the proper insecticide. S. O. S., SEEK OUT SOURCE! To locate ant nests, determine movement patterns and the worker ants will often lead you back to the nest. Another good technique is to use small bait stations to trick the ants into revealing their nest locations. Use soft drink or pill bottle caps baited with small amounts of peanut butter, jelly and bacon grease; one or more of these three foods will attract any ants. Watch the ants as they locate this food and begin taking it back to the nest. They may even establish an odor trail which other workers will use to come to the food source, so an ant column may develop. When you locate Ant City, USA, treat it with Baygon, Sevin, Dursban, Diazinon or a Boric Acid Bait; the species of ant will determine which you use. Follow label directions.

Pyramid Ant larva and worker
(VWR)

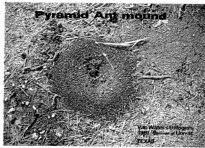

Pyramid Ant mound
(VWR)

Little Black Ant worker
(VWR)

Thief Ant worker
(VWR)

LITTLE FIRE ANT

The use of modern chemicals reduced Little Fire Ant populations to the point they were no longer considered a menace in Florida, but in recent years the reduction in use of these chemicals has allowed the populations to increase so they have again become a serious problem in some areas. In infested citrus groves, the ants are predominantly found on the leaves, around fruit and on the trunks of the citrus trees. This tiny creature is noted for its painful and long-lasting sting. Recently, pickers working in an infested lime grove in Hendry County demanded double wages for picking the limes.

Little Fire Ants are about 1/16 inch long and are brown to golden-brown in color. They are unusual in that they have no definite nesting areas. Have you heard of "squatter's rights"? These critters practice it. They may nest under leaf debris, rotten limbs, stones, clumps of grass or in the crotches of trees. Nests are frequently found behind the sheath of palms or palmettos. The species is highly adaptable in that it will nest in both open and shaded situations, seeming to thrive equally well under moist or dry conditions. One should not consider the ants contained in a nest as a solitary colony; a single "nest" will contain several reproducing queens, numerous workers, pupae, larvae and eggs. Nests within an area are obviously connected by worker movement. Little Fire Ants require tropical or subtropical conditions, so they are more a south Florida problem.

In homes the Little Fire Ant may infest clothing, beds, furniture or food; fats, peanut butter and other oily materials seem to be preferred. In nature, they are prone to tend honeydew-secreting insects such as aphids, scales and mealybugs.

Little Fire Ant = Order: Hymenoptera; Family: Formicidae

CONTROL CLUE

Guess what! There are no insecticides specifically registered for the Little Fire Ant. Call your local extension agent for further information.

P.S. If I had a Little Fire Ant problem, I think I would spray Malathion or Diazinon and ask questions later.

Little Fire Ant illustration (Photographer wanted extra pay for the real thing; we couldn't afford it!)
(DPI)

PHARAOH ANT

The Pharaoh Ant is perhaps the most difficult household insect pest to control. They are quite small (less than 1/16 inch long) and are a light tawny-brown color. Pharaoh ants adapt well to living in homes because they can readily nest in wall voids, appliances, linens, heating ducts, light fixtures and attics. They prefer to be close to heat and moisture sources.

Pharaoh ants are commonly called "sugar ants," but I promise you their diet is more diversified. They will have a family reunion on a drop of bacon grease overlooked on the kitchen range. Draw your bath in the morning and find a hundred and fifty drinking water on the shower head. Brush your teeth and see that interminable trail coming from behind the mirror heading for lunch on your toothpaste tube. Put on the shirt you wore to dinner last evening and find them crawling pell-mell up the sleeves heading for that spot of salad dressing inadvertently splattered on your collar. Those cookies for the office left on the breakfast table will be the occasion for an orgy that would have awed the Romans. Welcome to pharaoh ants!

This ant is a major pest not only of residences, but also of commercial bakeries, factories, office buildings, apartments . . . and also hospitals. Pharaoh ants have been found on burn victims and newborn babies, subjecting both to increased risk, because this ant has been known to transmit such pathogens as Salmonella, Staphylococcus and Streptococcus.

Pharaoh Ant = Order: Hymenoptera; Family: Formicidae

CONTROL CLUE

Pharaoh ants have the ability to survive most conventional household pest control treatments. Careless application of insecticides often makes the situation worse. Even an intensive, properly conducted control program may require two to three months to eliminate pharaoh ants. Do yourself a favor; call a good professional exterminator and save your sanity. What can the exterminator do that you can't? Use restricted insecticides and be more expert in their application. Believe me, you will need all the help you can get.

Pharaoh Ant workers tending eggs
(VWR)

Pharaoh Ant workers gathering food
(VWR)

_____Antlions (Doodlebugs)_____

When you were a child, did you ever play with a Doodlebug? Every Florida kid must know this critter in order to experience a normal, healthy childhood. If you don't know about Doodlebugs, you must be from "out-of-town." Let's proceed with your education — it's never too late to do the right thing.

A Doodlebug is an Antlion and is so named because the larvae of this family have strange eating habits. Adults resemble miniature damselflies. Female antlions lay eggs on the ground in sandy soil. When an egg hatches, the larva digs a pit shaped like an inverted cone with unbelievably smooth interior sides, and this larva lives almost completely buried at the bottom (the tip of the cone). This abode is a superbly designed trap, and should an ant or other small insect tumble in, it is seized in the powerful jaws of the larva and is sucked dry. Touch the upper edge of this sandy-sided trap with a twig and watch the action in the bottom of the hole. This guy does the original "jitter-bug," causing a sand-slide which will carry the prey into those waiting jaws. As the larva matures, it builds a silken cocoon in which it pupates. Doodlebugs are harmless (except to the critter that falls into the hole). Incidentally, even momma and

15

poppa antlions do good stuff; they eat some of the bad guys like caterpillars and aphids. If you don't have doodlebugs . . . get some.

Note to Mothers of Small Children: If you want a free baby-sitter for an hour, take li'l darlin' out in the back yard and find a doodlebug house. Of course, baby will drive that doodlebug crazy, but that's all part of being a doodlebug.

Antlions = Order: Neuroptera; Family: Myrmeleontidae

_____ **BENEFICIAL** _____

Antlion (Doodlebug)
(USDA)

_____ Aphids _____

When your young and tender garden plants begin to look puny for no apparent reason, you'd better begin checking for Aphids. Aphids are small, sluggish, soft-bodied insects often called Plant Lice. Sometimes they are referred to as Ant Cows and I'll tell you why in just a minute.

Aphids come in various colors, like brown, yellow, pink, black and various shades in between. These fellows are suckers in the truest sense of the word. There are numerous species of these insects and hardly any plant escapes their attack.You will find them on trees . . . on bushes . . . on potted plants, congregating on the new tender growth, sucking plant sap, causing stunting and leaf curling and leaving that most undesirable honeydew deposit. This sticky honeydew can often be found on cars parked beneath heavily infested trees. If periodically "mildew" gives your car fits, you had better check the tree for aphids or one of the other honeydew-excreting insects. (And wash your car more frequently.)

Large numbers of aphids may cause plants to gradually wilt, turn yellow or brown and die. They are the most important agents in the dissemination of plant virus diseases, and only a brief period of feeding by a single infected aphid may infect and eventually kill a plant.

Certain ants caress aphids, inducing them to produce more honeydew. The ants eat the honeydew. If aphids become crowded on a plant, the ants will often move some aphids to uninfested plants in order to supply them with a more ample food supply. They will tend them like a herd of cattle . . . hence, the name Ant Cows. Incidentally, aphids give birth to living young and breed prolifically. Some are winged and will fly.

Aphids = Order: Homoptera; Family: Aphididae

CONTROL CLUE

Some gardeners simply wash them off a plant with the garden hose, but better you should use a shot of Cygon, Malathion or Diazinon. Spray plant surfaces thoroughly, especially the undersides of leaves and young developing buds. If you prefer to use one of those insecticidal soaps, do it. Fortunately, aphids aren't difficult to control.

Aphids (Plant Lice)
(USDA)

Aphids tended by ants
(USDA)

Aphids and predator
(USDA)

Spider predator feeding on aphid
(VWR)

Giant Bark Aphid
(DPI)

Cabbage Aphids on cabbage
(VWR)

Pea Aphid stem mother giving birth
(VWR)

Black Citrus Aphids on orange twig
(VWR)

Armyworms

To be invaded by a horde of Armyworms is to know how Custer felt at Little Big Horn. Armyworms get their name from their feeding habits. They seem to appear all at once, eat everything, and then more or less disappear all at once. Armyworms are caterpillars that feed throughout the daylight hours.

There are four major species in Florida that are distinguished as follows:

Fall Armyworm — varies in color from light tan to green to black with three yellowish-white hair lines down the back from head to tail. On each side next to the yellow lines is a wider dark stripe and next to it an equally wide, somewhat wavy yellow stripe blotched with red.

Beet Armyworm — is green with prominent dark lateral stripes and a small black spot located on its side just above the second pair of true legs. (The true legs are the first three pairs right after the head.)

Southern Armyworm — is dark grey to nearly black and is marked with yellow stripes on its sides. This caterpillar is also considered a climbing cutworm.

Yellow-striped Armyworm — has a pair of triangular black spots on most of its body segments and often has a bright orange stripe just outside these spots on each side.

The fall armyworm is generally the most prevalent in Florida. Its name should not lead you to believe that it is a fall-of-the-year species. This caterpillar appears each spring and continues with new generations every 30 to 40 days. Feeding continues into fall. This insect cannot over-winter in any section where the ground freezes and must migrate from warm winter areas every spring.

One year in south-central Texas, I saw armyworms so bad you could actually hear 'em. And if I didn't hear 'em, I thought I did. Sounded like cottonwood leaves rustling in a soft breeze.

Armyworms = Order: Lepidoptera; Family: Noctuidae

CONTROL CLUE

In your garden use Sevin . . . and hurry! Armyworms are fairly easy to control with most recommended insecticides with the exception of the beet armyworm. This fellow must be properly identified and specific recommended control measures taken preferably before he reaches ½ inch in length.

Fall Armyworm having lunch
(USDA)

Yellow-striped Armyworm who already had a bite
(USDA)

Beet Armyworms about to have lunch
(TAEX)

Armyworm about to do it to a bald cypress
(DOF)

_____ Assassin Bugs _____

Assassin Bugs come in several different models. They are sometimes brightly colored, sometimes less so. They present an ominous manner, but are actually beneficial. Assassin bugs are very similar in appearance to leaffooted bugs and the adults are winged. Those "cocked" front legs prompt some to describe this critter as a "funny-looking praying mantid."

If ever you are invited to lunch by one of these friendlies . . . don't go. A favorite meal consists of stabbing into a big ol' caterpillar and then proceeding to suck the body juices from it. They probably do this same number on other insects also. WOW! What a way to live, but to each his own, I guess.

Assassin Bug = Order: Hemiptera; Family: Reduviidae

_____ **BENEFICIAL** _____

Assassin Bug adult
(VWR)

Assassin Bug having caterpillar lunch
(TAEX)

_____ **Bagworms** _____

Bagworms occur in Florida, but frankly they are no big deal. They are a "sometime" problem. But let's talk about 'em anyway; this chapter might win you some points in Trivial Pursuit. Bagworms are usually found on arborvitae and juniper, but also will prey on maple, box elder, eucalyptus, sycamore, willow, black locust, elm, basswood, poplar, oak, apple, cypress, wild cherry, sassafras, persimmon and cedar. WOW, what a menu! (If these critters *were* a major problem, half the trees in Florida could be at risk.)

Bagworms have the unusual habit of spending their entire larval stage inside a silken bag. It is very strong and well camouflaged with bits of twigs and leaves from the host plant. The early stages of an infestation are difficult to detect. As the larvae grow, the most obvious evidence of their presence is the defoliation of the upper portion of the tree and the presence of those Christmas-tree-like ornaments attached to the twigs and leaves where larvae are feeding.

Another unusual trait of the bagworm is that the adult female never leaves her bag. She is grub-like in appearance, has no wings or eyes and has no functional legs, antennae or mouth parts. Her body is soft, creamy-white and is nearly hairless. (No wonder she stays in the bag; how would you like to be married to that.)

Bagworms = Order: Lepidoptera; Family: Psychidae

Bagworm on eucalyptus
(DOF)

Bagworm having a coming-out party
(DOF)

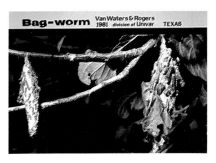

Bagworm cocoons on plum tree
(VWR)

Natural controls play an important role in keeping populations at tolerable levels. Bagworms may be killed by a variety of wasp parasites, by birds that prey on the young larvae and by low winter temperatures. Sprays should be applied when young larvae first appear in the spring. Cygon, Orthene or Sevin are good sprays. Read the labels.

On small yard and ornamental plantings, a simple method of control is to hand pick the bags and burn them immediately. Don't leave a bucket of bagworms out in the garage to be disposed of the next day. Those guys will crawl out of that bucket overnight, and you will be re-picking bagworms off of your car, the walls, the ceiling and everything else inside your garage. Listen to this voice of experience.

Bark Louse, Psocids

The Bark Louse prompts more questions than the Tooth Fairy. Every year, beginning about late spring, on-the-air gardening experts receive predictable calls and letters from worried fans asking about a mysterious, silvery web that covers the trunk and limbs of a favorite tree. This phenomenon occurs suddenly, frequently overnight; you walk outside one morning to get the paper and there it is, that eerie white web, coating the tree. Favorite Florida host trees are rough-barked hardwoods and palms. Surely this is the work of an evil force from outer space, or an insidious plague that will inevitably destroy the tree!

Don't worry. This is the work of the bark louse who, with his family, has come to do you a favor. These tiny creatures are scouring the bark for fungi, spores, pollen, lichen and other debris which they eat. While performing this housekeeping chore, they spin this web for protection. Usually the web will disintegrate in two or three days and will disappear almost as suddenly as it appeared. No control is necessary. You could blow the web away with water pressure from your garden hose, but don't you dare. These little guys have it tough enough in trying to clean up that tree trunk for you.

Bark Louse, Psocids = Order: Psocoptera; Family: Psocidae

_____ **BENEFICIAL** _____

Bark Louse on tree bark
(VWR)

That eerie, white web of Bark Lice
(DOF)

Bark Lice webbing close-up
(DOF)

_____ Bees _____

HONEYBEE

The Honeybee is everybody's friend. She is wrongfully blamed for many stings that are actually performed by the aggressive "yellow jacket" paper wasp. This popular insect is the prime practitioner in the insect world of the old philosophy . . . Live and Let Live.

Honeybees make honey. You know that. But do you know the major contribution the honeybees make to mankind is the function of plant pollinator? This industrious critter is definitely beneficial, but she will pop it (stinger) to you if you mess with her. Don't fool around her house (hive), and woe be unto your bottom if you sit on her. Every summer I lift a number of honeybees out of our swimming pool with my bare hand. But don't try this with your thumb and forefinger. This treatment is a squeeze and she'll sting you. Lift her out with the palm of your hand, and dump her out on the grass. Honeybee will buzz her wings several times to dry them and off she will fly to do her chores, and you will have done your good deed for the day. If you're out in the garden and one buzzes around your face, don't swat at her. She doesn't deserve that. Stand perfectly still. She will satisfy her curiosity and fly away. If perchance she sits on your nose, for god's sake, don't slap her . . . and you know why.

So, now you have a few tips on how to get along with Honeybee. Remember, be nice. Live and Let Live. And eat more honey!

Honeybee = Order: Hymenoptera; Family: Apidae

_____ **BENEFICIAL** _____

Honeybee adult
(VWR)

25

WILD BEES

Honeybees have some kinfolks. There are several species, but let's save time by referring to them simply by the Wild Bees. Adult colors will vary, but blacks, oranges and yellows are the most common. Some are metallic green or blue.

Wild bees nest in a variety of locations. By the way, not all bees live in colonies; many live alone. BUMBLE BEES nest in the ground and their life-style is simple; they won't bother you if you don't bother them. CARPENTER BEES are common in Florida and look very much like bumble bees. but they nest differently. Carpenter bee activity is commonly found around wooden fences, patio covers, wood shingles, roof eaves, porch ceilings, doors and windowsills. In establishing a nesting site, the female bores into wood at a right angle to the surface. The characteristic entrance tunnel is clean-cut, ½ to 1 inch deep, and is approximately ½ inch wide — about the size of a dime. Approach the carpenter bee next and you might find a bee "helicopter" hovering over your head. This is the male doing his duty. Although he is frightening because of his loud buzzing and large size, the male does not sting. The female will sting only if handled, so don't mess with her. Damage to structures is usually slight and amounts to cosmetic defacement, but try to sell that line to your wife when she finds that 45-caliber "bullet-hole" out by the back door.

The phantom-like LEAFCUTTING BEE is blamed for stealing those leaf hunks off your rose bush. My friend Baxter Williams insists (in jest) that the leafcutting bee is merely the figment of entomologists' imagination and doesn't really exist. Baxter contends he has never seen a leafcutting bee at work. Not many people have. This bee is indeed a solitary critter who nests in the ground and uses those leaf hunks to line its nest.

Faults notwithstanding, wild bees also do an excellent job of pollinating. Some native bees are better pollinators than honeybees. Credit where credit is due, folks. Let's have a round of applause for the Wild Bees.

Carpenter Bee = Order: Hymenoptera; Family: Anthophoridae
Bumble Bee = Order: Hymenoptera; Family: Apidae
Leafcutting Bee = Order: Hymenoptera; Family: Megachilidae

CONTROL CLUE

Remember, these Wild Bees are beneficial, but if you are pushed to a control for the carpenter bee, use liquid Sevin applied into the nest entrance and on a wide area of wood surface around the entrance hole. Wait 12 to 24 hours after application, then plug the hole. New adults will continue to emerge if insecticides are not used before plugging. Take care. Females are very aggressive, particularly during nesting activity. Bumble bees are treated more often than carpenter bees. Treat a bumble bee's nest by pouring the insecticide down the entrance hole, but remember they have a front door and a back door, so be sure you plug one before you treat the other or you're gonna get a painful surprise.

Bumble Bee adult
(VWR)

Valley Carpenter Bee/Bumble Bee comparison
(VWR)

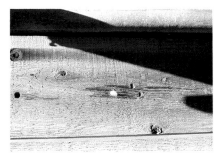

Carpenter Bee holes in cedar fence
(JEE)

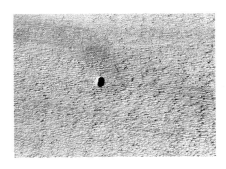

Carpenter Bee entry hole close-up
(JEE)

Leafcutting Bee cutting leaf
(TAEX)

___Beetles_____

AMBROSIA BEETLES

Ambrosia Beetles and a well known comedian have something in common . . . They Don't Get No Respect! Here's what I mean. Ambrosia Beetles generally utilize "down" wood like seasoned logs or lumber, dead trees *or dying or even severely stressed trees.* If you find Ambrosia Beetles in a seemingly healthy tree, that tree is not as healthy as you think. And if the tree ultimately dies, blaming Ambrosia Beetles for its demise is rather like blaming an undertaker for . . . death.

Adult Ambrosia Beetles emerge from infested material year-round. The mamas deposit eggs in galleries that have been constructed in the sapwood and heartwood of a host hardwood or softwood tree. After hatching, the larvae or immature beetles feed on a special fungi — the ambrosia — that has been introduced by the parents. Ambrosia means "nectar of the gods," but this "nectar of the beetles" is hardly that. The fungi feed on the wood fibers and the larvae then feed on the fungi. These fungi cause that characteristic blue-black stain on the gallery wood. Mating of adults may occur in the galleries after which the females fly to new hosts, or both sexes may fly to a new host and then mate. There are several generations per year. (Atta' boy, beetles!)

Infestation signs may be one or more piles of white, fluffy boring dust at the tree base; pinholes or shotholes through the bark and into the sapwood and heartwood; split wood with small diameter galleries running with and against the wood grain; open galleries, free of frass or sawdust; and that wood stain caused by the ambrosial fungi.

Ambrosia Beetles = Order: Coleoptera; Family: Scolytidae

_____ CONTROL CLUE _____

Cut the infested tree and haul it away . . . the darn thing is gonna' die anyway. I wouldn't even keep it for firewood. If you're worried about neighbor trees, drench the bark areas of those trees with Lindane or Dursban; with a bit of luck, that might deter further infestations. Don't forget your local county agent or forester. Ask for help.

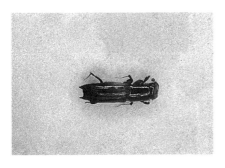

Ambrosia Beetle boring dust at base of slash pine at Gulf Island National Seashore, Pensacola, Florida
(DOF)

Ambrosia Beetle adult
(SFI)

Ambrosia Beetle boring dust at base of oak tree
(DOF)

Sap-stained Ambrosia Beetle hole in India rosewood
(DOF)

Cadillac Beetle entrance holes in pine
(DOF)

Ambrosia Beetle larva, pupa and adult
(VWR)

BLISTER BEETLES

Several species of Blister Beetles do business in Florida. The adults are rather soft-bodied, long, slender and long-legged with distinct body divisions. They may be black, gray, striped or even brightly colored.The family is most unusual in that the larvae of many species are beneficial, but the adults are injurious. These fellows aren't all bad; just mostly bad. Adults of most species feed on flowers or foliage; some are pests on potatoes, tomatoes, beets or other vegetables. All larvae seem to be specialized predators; some feast on pods of grasshopper eggs, but others will wait on flowers for a visiting bee, attach to it, and be carried to the bee's nest where it remains in the cell and eats the bee eggs and the stores provided for the bee larvae. If there is a heavy infestation of blister bugs in a flower patch, bees tend to stay away . . . bees aren't dummies. Blister beetle eggs are laid in masses in the ground or under stones or on the food plants of adults.

When disturbed, blister beetles exhibit "reflex bleeding" — they emit blood from knee joints and other parts of their body. The blood contains a peculiar, odorless chemical, cantharidin, which causes blisters on human skin. (Guess how the blister beetle got its name!) Taken internally or absorbed through the skin, cantharidin is highly toxic to mammals. In Jacksonville, Florida, thoroughbred horses worth over $200,000 died from eating locally grown alfalfa hay that was infested with blister beetles. Cantharidin is still used in the United States by dermatologists as a wart remover. Dried, pulverized bodies of adult blister beetles, containing cantharidin, is the source of "Spanish fly" . . . and before you write for the recipe, be advised, I don't have it. (I do make great home-made bread and chicken soup, however.)

Blister Beetles = Order: Coleoptera; Family: Meloidae

CONTROL CLUE

Like for other beetles — Sevin or Diazinon. In small gardens, it may be sufficient to simply pick 'em . . . but use throwaway gloves.

Striped Blister Beetle
(USDA)

Blister Beetle (black)
(VWR)

Margined Blister Beetle
(USDA)

COLORADO POTATO BEETLE

Colorado Potato Beetle adults are husky, yellow and black-striped beetles that are about ⅜ inch long. These critters are also known as Potato Bugs, but will often devastate tomato, eggplant and pepper plants as well. About 500 eggs are deposited in batches of 24 or so on the underside of leaves. The eggs hatch in 4 to 9 days and the larvae become full grown in 2 to 3 weeks. Larvae are reddish in color, and are humped; they also are equipped with two rows of black dots on each side of the body. Multi-generations per season occur in Florida.

This pest was once a bane of commercial potato growers and still causes considerable damage in home gardens. Both adults and larvae feed on leaves and can completely strip a plant in short order.

Colorado Potato Beetle = Order: Coleoptera; Family: Chrysomelidae

_____ **CONTROL CLUE** _____

Use a dust formulation of Sevin or Diazinon when these insects and damage are first noticed. Repeat every week till the infestation is eliminated.

Colorado Potato Beetle adult
(USDA)

Colorado Potato Beetle larva
(USDA)

COTTONWOOD LEAF BEETLE

Cottonwood Leaf Beetle can be tough on the foliage of a favorite shade tree. Adults become active in the spring and feed on tender foliage and bark. After mating, the females lay eggs on the underside of the leaves. Young larvae emerge and feed in groups, skeletonizing the leaves; older larvae feed separately and consume all but the major leaf veins. Cottonwoods, poplars and willows are among the favorite hosts. There may be 3 or 4 generations per year and they may occur in large numbers on individual trees. By late summer, the whole family — larvae, pupae and adults — can be found on one tree. This is a family that believes in togetherness.

Adults are about ¼ inch long, have black heads along with rather colorful orange-red and gold wing covers that bear 7 purple-black irregular spots. Larvae are black when young, but later turn a dirty yellow with black legs and brown heads. Frankly, they look like — bird droppings!

Cottonwood Leaf Beetle = Order: Coleoptera; Family: Chrysomelidae

_____ **CONTROL CLUE** _____

Although infested trees seldom die from cottonwood leaf beetle defoliation, they suffer intense stress because of it. Predators, parasites, disease and the weather usually keep leaf beetle populations at low levels, but if you suffer a flare-up, spray with Sevin or maybe Orthene.

NOTE: The larvae emit a foul-smelling fluid when handled or disturbed. In the event you get the bright idea to handpick 'em, be sure you wear throwaway gloves, lest you be shunned by your family when you go back in the house.

Cottonwood Leaf Beetle larvae on a cottonwood leaf
(DOF)

CUCUMBER BEETLES

There are several species of cucumber beetles that attack field and vegetable crops. We will take a look at three: Spotted Cucumber Beetle, Striped Cucumber Beetle and Banded Cucumber Beetle. This family of critters reminds me of Ma Barker and her desperado sons. Mrs. Spotted, the most important pest of the group, must also bear the blame for the mischief of her progeny, Southern Corn Rootworms.

Cucumber beetle adults are all guilty of the same sins when it comes to goofing up your garden . . . like, they eat plants. Some favorites are cucumbers, squash, pumpkins, corn, lima beans and peas. They will often feed on stems and young leaves of newly-emerged seedlings. Their menu is enormous . . . over 200 common crops, grasses and weeds. Also, while gadding about your garden, they will spread serious diseases: for example, mosaic and bacterial wilt on your cucurbits. (Note: cucurbits are a group of vegetables . . . not any part of your anatomy.) And if you think this is bad, wait 'till you hear what the kids (rootworms) do.

Are you in tears? I really believe we should continue this grizzly tale down the line with the chapter on Southern Corn Rootworms, lest I completely ruin your day. Be brave.

Cucumber Beetles = Order: Coleoptera; Family: Chrysomelidae

CONTROL CLUE

Use Sevin or Diazinon. Treat at first sign of the beetles. Repeat at weekly intervals as plants become reinfested. Don't tarry! When you control cucumber beetles, you are also controlling rootworms.

Striped Cucumber Beetle and feeding damage
(USDA)

Twelve-spotted Cucumber Beetle
(VWR)

CYPRESS LEAF BEETLE

Cypress Leaf Beetles attack the foliage of Bald Cypress, Pond Cypress and Sand Pine. (Sand pine? Yes! These critters don't read entomology books, I guess, and don't know where they belong.) Adult beetles are common in the tree crowns by midsummer and outbreaks commonly occur over large areas. Damage typically becomes noticeable within a short period of time because congregations tend to be massive. They will concentrate on a single tree or a small group of trees and just as abruptly leave for another area of host trees. Meanwhile, you see tree foliage discoloring from green to a bright-to-dark-red and you stand there with your mouth open, wondering what happened to your poor tree.

Upon closer examination you will notice small linear gouges in the needles. Drought or other stress factors can bring on a similar needle color change, but none will make the small "slits." Infested trees shed those damaged needles, and within a few months it will be difficult to identify the affected area because of the new foliage growth. Another symptom of infestation will be a swarm of small yellowish beetles highly active around the tree crowns. The head and body of these adults are colored a pale yellow to gold with a dark stripe down the center of the back; wing edges are black. Larvae don't participate in this adult carnage and are thought to feed on the roots of grasses and weeds.

Cypress Leaf Beetle = Order: Coleoptera; Family: Chrysomelidae

---

CONTROL CLUE

The rather rapid refoliation by defoliated trees might suggest chemical control is not warranted in most cases, but if you just can't stand it, check with your area forester for a suggested control measure. I would be disposed to hit an affected tree with Sevin or Diazinon or Orthene, but another problem will be getting any spray material up to the top of the tree where the bugs are. Have you got a high-volume, high-pressure sprayer? This is getting too complicated! Better you find something else to worry about. Forget it . . . let the tree handle it.

Cypress Leaf Beetle adult
(DOF)

Cypress Leaf Beetle feeding
damage to bald cypress
(DOF)

ELM LEAF BEETLE

Elm Leaf Beetles can be found wherever elms grow, and 2 to 4 generations of this pest may occur per year. Adult beetles eat holes in the leaves, but the larvae cause most of the damage by skeletonizing the lower leaf surface. If large populations develop in residential areas, the beetles may become a further nuisance as winter approaches, because they will invade homes.

These beetles will begin to leave their protected overwintering sites in the spring in order to fly to nearby elm trees at about the time elm leaf buds begin to swell. Adults feed by chewing holes in the unfolding leaves. Egg laying begins a short time later. Masses of up to 25 yellow-orange eggs are deposited on the underside of leaves and these eggs will hatch in about a week. The larvae will then feed on the undersurface of the leaves, leaving only the veins and the upper leaf surface. Those damaged leaves soon dry, turn brown and may drop from the tree. When these conditions are severe and trees are without leaves for several consecutive years, limbs or perhaps the entire tree may die. Adult beetles are about ¼ inch long; larvae (worms) are about ½ inch in length.

Elm Leaf Beetle = Order: Coleoptera; Family: Chrysomelidae

Elm Leaf Beetle adult
(USDA)

Elm Leaf Beetle larva
(USDA)

FLEA BEETLES

Flea Beetles are some of the most annoying of all vegetable pests. As
beetles go, these guys are runts. Adults are approximately 1/16 inch long
and are equipped with enlarged flea-like hind legs. As a matter of fact,
flea beetles jump like fleas, but are not related to them. Some are striped,
but most are either black, brown or green. Larvae are small, slender and
white with a black band, and they have three pairs of legs.

Flea beetles attack numerous plants, but preferred vegetable hosts
are tomatoes, peas, cabbage, carrots, eggplants, turnips, mustard,
potatoes, beans and others. Both adults and larvae cause plant damage.

Adults chew numerous tiny, rounded or irregular holes in foliage so that leaves appear to have been peppered with fine shot. When feeding damage is heavy and there are many holes, leaves may wilt and turn brown; the host plant may become stunted and may even die. The larvae feed on roots or tubers. Life cycle from egg to adult may be completed in 6 weeks or less. One to four generations per year may develop.

Flea Beetle = Order: Coleoptera; Family: Chrysomelidae

_____ CONTROL CLUE _____

Don't let these midgets fake you out. They are easily over-looked because they are so small. Watch for those shot-holes in the leaves and then get after them quickly with a spray containing Sevin, Methoxychlor or Diazinon.

Flea Beetle adult on radish leaf
(VWR)

Flea Beetle damage
(TAEX)

GROUND BEETLES

Several species of Ground Beetles prey on a variety of other insects, insect eggs and other anthropods (critters). One species, the Seed Corn Beetle, will occasionally damage germinating seeds or seedlings, but this is an exception. Considering their many other attributes, we might readily score them 98% beneficial.

Adults are normally dull to shiny black. Some species may be brown or may have red legs, or may have iridescent green-blue elytra (wing covers). They are very active and are fast movers. Some are quite large . . . up to one inch in length.

Larvae are fleshy grubs with large jaws. They generally remain in the soil, so you're not likely to see them. If you accidentally found one, I think you would kill it, because it is such an ugly thing. Shame, shame on you.

Gound Beetles = Order: Coleoptera; Family: Carabidae

Ground Beetle (Fiery hunter)
(USDA)

Ground Beetle (Caterpillar hunter)
(VWR)

Tiger Beetle
(USDA)

39

LADYBIRD BEETLES

Here's the darling of the insect world, commonly called "Lady Bug." A real sweetheart! Typically colored red with black spots, she may also be black or black with red spots or even gray with black spots. Ladybird and her children (larvae) use chewing mouthparts to feed on other small insects that plague your garden. Oftentimes you'll find a bunch of them devouring plump, green aphids.

Learn her markings well lest you confuse her with the Mexican Bean Beetle or the Squash Beetle or one of the other bad guys and in error wipe her out. Mother Nature cursed Ladybird in another way . . . gave her UGLY children. You will wonder, can anything that ugly be good? Oh, well, you can't have everything.

Ladybird Beetle = Order: Coleoptera; Family: Coccinellidae

_____ **BENEFICIAL** _____

Twice-stabbed Ladybird Beetle
(DPI)

Orange Ladybird Beetle adult
(USDA)

Black Ladybird Beetle after pine scale
(VWR)

Ladybird Beetle larva attacking aphids
(USDA)

MEXICAN BEAN BEETLES

I guess the Mexican Bean Beetle is how Mexico got even with the United States for the Mexican-American War. The Mexican bean beetle looks very much like the ladybird beetle. They are relatives, but in epicurean habits the similarity ceases.

The Mexican bean beetle has sixteen black spots on its back. Its basic color is usually copper-like or orange. Size is like that of the ladybird, about ¼ inch long. Orange to yellow soft-bodied grubs about ⅓ of an inch long with black-tipped spines on their back may also be present. These are the larvae (babies). These critters primarily work on the underside of your bean leaves. The leaves appear to dry up and the plant may die.

You're likely to tolerate the Mexican bean beetle thinking it to be a ladybird, or you will wipe out the ladybird thinking it's the Mexican bean beetle. Pay attention.

Mexican Bean Beetle = Order: Coleoptera; Family: Coccinellidae

---

_____ **CONTROL CLUE** _____

Diazinon or Sevin. Be sure to apply to the underside of the leaf and be sure you have the right critter. The ladybird beetle is a beneficial insect and should not be eliminated.

Mexican Bean Beetle family
(USDA)

PINE BARK BEETLES

Pine Bark Beetles account for more damage to southern pine timber than any other group of insects. A substantial portion of the 3 billion cubic feet of timber lost in the South (folks, that amounts to a lot of homes and newspapers) is attributed to the BLACK TURPENTINE BEETLE (BTB), three IPS ENGRAVER BEETLES (IEB) and the SOUTHERN PINE BEETLE (SPB). In Florida, the most important are the IEB, followed by the BTB. Significant timber losses due to the SPB are rather infrequent in Florida.

Although different in many ways, this group of gangsters shares a common habit . . . they tunnel into and feed within the soft inner bark of pine trees. Trees struck by lightning or cars, damaged by fire, hurricanes, dogs, children, herbicide, construction, and on and on, are attacked first by bark beetles. Generally, adult bark beetles are first attracted to weakened or otherwise stressed trees, but during major outbreaks, seemingly healthy trees will be attacked also. It is not uncommon for two or more pine bark beetle species to infest a tree simultaneously. The tunneling of the larvae and adults severs the tree's nutrient transport system; equally damaging is the plugging of the water transport system by bluestain fungi which are introduced by the adults.

To state the pine bark beetle story simply: adults enter tree and tunnel egg galleries, eggs hatch, larvae and adults eat, larvae pupate, new bark beetles emerge to begin the cycle again in another susceptible tree. There can be up to 10 generations per year for the smaller IEB species, but only 2½ to 3 generations per year of the much larger BTB.

The foliage of severely infested trees changes from healthy dark green to light green, then to yellow, then to sorrel, and finally to reddish-brown and goodbye pine tree. If trees are infested in late fall, they may not fade until the following spring, but once the foliage change appears, kiss your tree goodbye.

Specific infestation signs: IEB . . . small lumps of reddish-orange pitch on tree stem or branches with reddish-orange boring dust and narrow Y-shaped or H-shaped galleries in the inner bark. BTB . . . large lumps of pitch on lower stem of tree with pitch tubes pinkish-white to reddish-brown that will fade to purplish-gray after a month or two and wide D-shaped galleries under bark. SPB . . . small lumps of whitish pitch on tree stem, reddish-brown boring dust and narrow S-shaped galleries under bark.

Black Turpentine Beetle = Order: Coleoptera; Family: Scolytidae
Ips Engraver Beetle = Order: Coleoptera; Family: Scolytidae
Southern Pine Beetle = Order: Coleoptera; Family: Scolytidae

Maintain healthy trees. Don't stress them by building sidewalks or patios over the roots or by digging excessively around the roots, etc. Some protection may be attained by drenching bark surfaces with Lindane or Dursban-4E. Infested trees should be cut and burned. Consult a Florida forester or county agent for more specific advice.

Ips Engraver Beetle adult
(DOF)

Black Turpentine Beetle adult
(DOF)

Southern Pine Beetle adult
(DOF)

SPB adult covered with mites
(SFI)

Temnochila virescens, predator of SPB
(SFI)

Ips Engraver Beetle pitch tubes
(DOF)

Black Turpentine Beetle pitch tubes
(DOF)

Ips Engraver Beetle pitch tube close-up
with trapped Ips beetle
(DOF)

Ips Engraver Beetle nuptial
chamber and egg galleries
under construction
(DOF)

Black Turpentine Beetle
parent gallery
(DOF)

Southern Pine Beetle adult in gallery
(SFI)

44

Clerid Beetles, predator of SPB
(SFI)

Southern Pine Beetle
galleries in loblolly pine
(DOF)

Another symptom of SPB
attack; boring dust in spider
web at tree base
(SFI)

Larva of the predator Clerid working
through the galleries bored by the SPB,
searching for prey
(SFI)

Adults of the 5 common species of bark
beetles that attack and kill pines. The
largest is the Black Turpentine Beetle;
next the Southern Pine Beetle; in the
middle is the Coarse Writing Pine
Engraver; then the Southern Pine
Engraver; then the Small Southern Pine
Engraver.
(SFI)

PINE COLASPIS BEETLES

Pine Colaspis Beetles seem to prefer slash pine as a host, but they have been found on many other southern pine trees. Feeding damage caused by large populations occasionally produces a spectacular browning effect of the needles similar to that caused by fire damage. Adult colaspis beetles chew the edges of needles to produce an irregular, saw-like edge which turns brown. Initially, only the tips of the needles show these signs of infestation, but later the entire needle may die, causing the whole tree to turn brown and appear to be dying. However, attacked trees do not die and little or no growth loss results.

This pest is sporadic in its occurrence and may not develop again in the same area for several years. Attacks usually occur in early summer, but by late summer the trees appear green and healthy again. The adult female pine colaspis beetle lays her eggs in the soil during the summer. Larvae hatch and feed on roots of grasses and other vegetation, and over-winter in this stage. The larvae pupate in the spring; adults emerge in early summer to feed. There is only one generation a year.

Pine Colaspis Beetle = Order: Coleoptera; Family: Chrysomelidae

---

CONTROL CLUE

Under forest conditions, no control measures are recommended for the pine colaspis beetle, but in your yard, practical control may be attained by using a Sevin spray.

NOTE: Pine Colaspis Beetles are night feeders, but so are Pine Chafers and Reproduction Weevils; detection and identification is often complicated by this circumstance. Just be sure you have the right critter in your sights before you pull the trigger. And here I go being honest again . . . as pine tree pests go in Florida, the pine colaspis beetle is "no big deal" at this time. Sorry I bothered you!

Pine Colaspis Beetle adult
(DOF)

Pine Colaspis Beetle feeding
damage to pine needle
(DOF)

Pine Colaspis Beetle and damage to
loblolly pine
(SFI)

SOFT-WINGED FLOWER BEETLE

Remember the red cross! And don't say I gave at the office. The red cross that I refer to might be a characteristic marking on the back of a Soft-Winged Flower Beetle. Several related species are found in gardens.This friend is about ¼ inch long and can usually be found on a bright, sunshiny day scurrying about on plants or flowers. This "Speedy Gonzales" preys on most any smaller insect. This critter's fast footwork enables it to cover so much plant territory, that he probably does a better job of pollinating than we give him credit for.

Soft-winged Flower Beetle = Order: Coleoptera; Family: Melyridae

A Soft-winged Flower Beetle
(VWR)

TORTOISE BEETLE

Tortoise Beetle adults are oval and flattened in shape and are about ¼ inch long. Colors may be golden, metallic blue to green or tan. The adults of several species are actually very striking in their metallic to iridescent colors.

Larvae are short and grub-like, and have a forked posterior appendage bent forward over the body. This fecal fork holds a case made of cast skins and excreta; it is held over the body much like a parasol. If a larva is threatened by an enemy, perhaps an ant, this case full of mess is pushed into the predator's face. (What a way to ruin an ant's day!) These critters get grade F in hygiene and look like animated dirt, similar to trash bugs who are also travelling dumpsters.

Both adults and larvae feed on the foliage of eggplant, sweet potato and other plants in the morning-glory and night shade family. These critters cut holes in leaves and sometimes will consume the entire leaf. A regular diner on Florida's sabal palm is a gun-metal blue tortoise beetle.

Tortoise Beetle = Order: Coleoptera; Family: Chrysomelidae

Palm Tortoise Beetle
(DOF)

Tortoise Beetle (gold bug), two species
(USDA)

TWIG GIRDLER BEETLE

Twig Girdlers are beetles that do dirty-work on your trees by pruning off limbs. They could well be arrested for Malicious Mischief, but there is motive to this wood-carving madness.

Female twig girdlers do this unauthorized pruning after laying eggs in the limb part that is ultimately severed. Girdling begins during late summer, but the damage may not be noticed until the fall when the wind causes the withered branch to snap off and fall to the ground. It's all part of the twig girdler master plan for procreation. The eggs hatch into larvae which grow in the fallen limb. The larvae ultimately emerge, resulting in more mamas which will in time cut off more limbs.

Twig girdlers will attack apple, ash, dogwood, cedar, elm, hickory, juniper, mimosa, oak, peach, pear, pecan and other shade trees.

Twig Girdler Beetle = Order: Coleoptera; Family: Cerambycidae

49

If you want to try chemicals, use a Lindane spray when girdling is first noticed.

I do have another remedy. (This one will offer no better control, but it will make you feel good.) Catch mama girdler in the act and if you can reach her, bash her with a club. Next, gather any cut limbs that may be on the ground and burn them; there will be fewer twig girdlers next year. True, the damage will have been done, but just think, all that gnawing for nothing. Fight dirty!

Twig Girdler and damage to limb
(USDA)

WHITEFRINGED BEETLE

The Whitefringed Beetle is a pest of crops grown in the western and northern part of Florida. It attacks over 350 species of plants including field and vegetable crops. The adult beetle is about ½ inch long and is gray in color with faint white strips on its sides. It causes damage by feeding on the leaves of plants, but the critical damage is caused by the grub-like legless larva. Grub damage is confined to plant parts below the soil surface, as this varmint does not climb . . . remember, no legs! The adult is unable to fly, so it goes to greener pastures by walking ¼ to ¾ mile per year or by being moved about by man. Don't you know whose feet hurt every night?

Whitefringed Beetle = Order: Coleoptera; Family: Curculionidae

---

_____ CONTROL CLUE _____

Sit down, I have bad news again. Control is almost impossible. Certain soil insecticides which provided the only successful treatment have been banned. Since the host is so numerous, crop rotation is of little value. Often it becomes necessary to simply abandon a piece of infected land. For the present, if you see a whitefringed beetle, hit it with a rock.

Whitefringed Beetle adult
(USDA)

WOOD DESTROYING BEETLES

Wood destroying beetles destroy wood. And if that wood is a piece of pine 2x4 out in your garage, that's one thing; but if the wood is an antique Louis XIV chair or a prized family heirloom, then that's quite another thing. Wood destroying beetles go by many names. Let's look at a few.

POWDERPOST BEETLES attack only seasoned hardwood and generally feed on seasoned wood. Thus, powderpost beetles found in hardwood floors or furniture would not be expected to also attack the softwood (e.g., pine) structural timbers in a home. On the other hand, DEATHWATCH BEETLES and FALSE POWDERPOST BEETLES can attack both softwoods and hardwoods and also generally feed on seasoned woods. One of the most significant wood infestors is the OLD HOUSE BORER, which generally attacks structural softwoods. Contrary to its name, it is often found in newer homes built with infested wood and will readily reinfest structural timbers. Wood that is improperly kiln-dried or treated, or wood that is stored too long is more likely to be attacked by wood borers.

There are several indications that wood boring beetles are present in a wood member. Immature beetles tunneling in wood may cause an audible rasping or ticking sound most often heard during quiet times at night. Another indication may be a blistered appearance on the wood caused by larvae (worms) tunneling just below the wood surface. In feeding, beetles often push white to yellow powdery frass (sawdust-like stuff) through holes made in infested wood. The consistency of the frass ranges from fine to coarse, depending on the species. Entry or exit holes without the frass may also be observed. Occasionally, wood staining or the actual sighting of adult beetles will be noted. Adult beetles emerging from their havens are often attracted to lights or windows.

The majority of wood borers (flatheaded borers, roundheaded borers, ambrosia beetles and timberworms) are found shortly after a structure is built. (This is especially true of new log homes or cabins.) Adults of these species generally will emerge within a few years after a building has been constructed and will not reinfest other wood. There are exceptions, of course, such as the old house borer. Proper identification is extremely important since controls vary according to species.

Powder Post Beetles = Order: Coleoptera; Family: Lyctidae
Deathwatch Beetles = Order:Coleoptera; Family: Anobiidae
Old House Borer = Order: Coleoptera; Family: Cerambycidae
Wharf Borer = Order: Coleoptera; Family: Oedemeridae
Southern Pine Sawyer = Order: Coleoptera; Family: Cerambycidae
Furniture Beetle = Order: Coleoptera; Family: Anobiidae

_____ CONTROL CLUE _____

Prevention is the best control. Carefully inspect antique furniture, picture frames, bamboo products and other wood items before buying them. Consider any evidence of emergence holes, larval infestation or frass presence to be an indication of active infestation. Store fireplace wood as far from home as possible and only bring in firewood that will be immediately burned. Adult beetles can emerge from wood stored in the home and infest structural wood or furniture.

Small home furnishings, wooden artifacts or small furniture pieces may be treated by freezing, but it is necessary to maintain the items at 0 degrees Fahrenheit for several weeks to effect control. A localized infestation may be treated by applying a residual insecticide such as water-emulsified Lindane. Two coats are desirable, with the second being applied before the first completely dries. Proper treatment of unfinished wood may be effective for 10 years or more, but this will eliminate only adults that emerge or that attempt to reinfest. Larval development beneath the wood surface often continues and deep-boring beetles such as the old house borer are usually not reached.

Fumigation is the most reliable and effective method of eliminating wood boring beetles, but because it is a costly, highly technical land hazardous process, *You must not try to do this.* Fumigation must be left to qualified pest control operators who are experienced in employing this technique.

Deathwatch Beetle adult (Anobiid)
(VWR)

Deathwatch Beetle larva and damage
(VWR)

Old House Borer adult
(SFI)

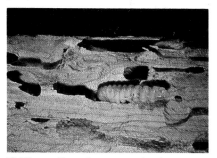
Old House Borer larva; note that frass has been removed from feeding tunnel.
(SFI)

53

Powderpost Beetle adult female with extruded ovipositor and eggs
(SFI)

Powderpost Beetle larvae; they often damage articles manufactured from hardwoods; i.e., flooring, ax handles, etc.
(SFI)

Wharf Borer adult; larvae of these beetles cause damage to moist or wet wood in contact with water or soil.
(SFI)

Furniture Beetle adult; these beetles are usually found damaging softwood structural timbers in the crawl space areas of homes.
(SFI)

Southern Pine Sawyer larvae; these beetles are pests of drying, bark-covered wood. Larvae feed into the wood and may cause damage in lumber cut from infected logs. The damage resembles that of the Old House Borer.
(SFI)

Southern Pine Sawyer adult; these are chiefly wood borers, but they bore in the inner-bark for some time before entering the wood.
(SFI)

Hunting Billbug

The Hunting Billbug, also known as the Zoysia Billbug, is a pest of various grasses in Florida. By far, the most commonly recorded host is Zoysia grass, but others are Bermuda grass, Pensacola Bahia grass, St. Augustine grass and centipide grass. Most of the damage results when lawn grasses are planted in areas which formerly had high populations on natural grasses and sedge cover. This seems to be particularly true in moist areas such as the muckland around Lake Okeechobee which is used for commercial sod production. Significant Hunting Billbug infestation leaves sod root systems so weakened that sod squares fall apart in handling.

This critter is a weevil. It is typically weevil-like in appearance with a short, fairly broad, recurved snout. Colors vary from gray to black with reddish or brown areas sometimes visible. Adults are also susceptible to a greasing from body oils which tend to give them a dirty appearance.

Eggs are laid in the grass leaf sheaths or top of the crown; they hatch in three to ten days, and the newly emerged larvae (ugly grubs) feed on the inner leaves as they work their way down to the roots.

Hunting Billbug = Order: Coleoptera; Family: Curculionidae

_____ CONTROL CLUE _____

Several insecticides are successful in controlling this pest, but I like Oftanol the best. Read the label.

Hunting Billbug adult and immatures
(IFAS)

Hunting Billbug larva and damage to sugar cane
(IFAS)

_____Borers_____

BLACK TWIG BORER

Black Twig Borer females (only females, mind you) emerge from small branches and twigs of host trees by late winter to early spring, and then bore into shoots to lay eggs. The males never leave home — they just die. (Who said it's a man's world!) The larvae feed on fungi growing on the tunnel walls; they pupate inside the damaged shoot and over-winter as adults. There may be several generations per year.

Common symptoms: foliage discoloration and/or wilting, branch dieback and small holes on the underside of twigs and branches. A wilt-type fungus usually causes a sunken canker in the area of the holes. Adult females have a small but stout cylindrical body and are light to dark reddish-brown in color. Males are rarely seen; you gotta' dig for 'em. Larvae will likely have dark brown heads, yellowish-white bodies and be legless.

First detected in Florida in 1941, the black twig borer is now distributed nearly statewide. Its aggressiveness in infesting healthy trees distinguishes it from most other Ambrosia Beetles. Common hosts are apple, boxelder, camphor, dogwood, elm, golden raintree, pecan, redbud, red maple, southern magnolia and many other hardwood trees.

Black Twig Borer = Order: Coleoptera; Family: Scolytidae

Black Twig Borer adult
(DOF)

Dogwood branch infested with Black Twig Borer
(DOF)

LESSER CORNSTALK BORER

The Lesser Cornstalk Borer attacks more than 60 known plant species in Florida. It is a particularly serious pest on corn, peanuts, soybeans, southern peas and various varieties of beans grown as vegetables. A sudden wilting of your garden, especially in August, should make you suspicious.

The female moth lays eggs on the host plant, but again the larva (baby) does the damage. The young caterpillar feeds first on the foliage and then begins to burrow into the stem. The worm attaches silken-like tubes to the stem of the plant, at or just below the soil surface, and then begins to tunnel. In order to find this rascal, the plant stem must be carefully cut into with a knife. Upon exposure or if prodded the worm will "jump around" violently.

In addition to doing the St. Vitus's dance, the worry-wart will be recognized by markings — bright green to turquoise, striped longitudinally with brown, with a dark brown to black head. It's a small critter, ¼ to ½ inch in length.

Less well known is the fact that seedlings of forest nurseries are susceptible to infestation. In 1981 nearly 1 million seedlings were killed by larvae of the lesser cornstalk borer in a forest nursery located in Central Florida.

Lesser Cornstalk Borer = Order: Lepidoptera; Family: Pyralidae

Lesser Cornstalk Borer
caterpillar
(DOF)

Lesser Cornstalk Borer
damage and galleries in
pearl millet
(DOF)

PEACH TREE BORER &
LESSER PEACH TREE BORER

These borers feed in the bark of peach trees; damaged trees seem more susceptible to an intrusion of these pests, but seemingly healthy trees will also be invaded.

Peach Tree Borers attack the trunk from about 6 inches above the soil to 3 inches below the soil. They begin life as eggs, then progress to larvae, then pupae, then finally to adult moths. In Florida, the major adult flight occurs in late summer and fall, with peak emergence in late September to early October. Eggs hatch into larvae in about 10 days, bore into the bark near the hatching site and begin feeding in the cambium, that vital area just under the bark. Thick, gummy sap usually oozes from the entry holes. And now your tree is in big trouble. If the cambium layer is damaged extensively, the tree will die. Severely infested trees often can be identified by the dead or dying limbs.

It is not known how many larvae it takes to cause economic damage. A single feeding worm can devastate a young tree that is 1 to 2 inches in diameter. A tree 10 to 12 inches in diameter may safely harbor 1 or 2 larvae. Any more should be treated.

Lesser Peach Tree Borers attack the upper tree trunk and scaffold limbs. Moths (adults) first emerge around March and continue at fairly constant levels through October; otherwise, life-style and damage is comparable to that of their kinfolks, the peach tree borers.

Peach Tree Borer = Order: Lepidoptera; Family: Sesiidae

_____ CONTROL CLUE _____

This critter is a tough cookie and will require more than incidental spraying. The specific timing of sprays for most effective borer control has not been determined for Florida, but try this recipe: Dursban applied top-side at apt times, with a Thiodan drench at the base of the tree and be sure to make one of the applications at the end of August. Remember these "no-no's": do not contaminate the fruit; do not spray within 14 days of harvest; do not deviate from label instructions.

You really need to know much more than this; why don't you call your local county agent.

Note: Don't waste your time with the old car-battery-hooked-to-tree cure. Won't work.

Peach Tree Borer female
(USDA)

Peach Tree Borer larva
(USDA)

SHADE TREE WOOD BORERS

The two predominant groups of Shade Tree Borers are called the Roundheaded and the Flatheaded Borers. Adult roundheaded borers are called Longhorned Beetles because of their long antennae. Adult flatheaded borers are called Metallic Wood-Boring Beetles because of their iridescent metallic luster. The cylindrical, hard-shelled, longhorned beetle varies from ¼ inch to over 3 inches in length. Markings range from contrasting colored bands, to spots or stripes. The antennae are usually from one-half to over one-and-a-half times the length of the body, depending on species. The metallic beetle usually is beautifully colored, oval shaped and ⅓ to 1 inch long.

The roundheaded group is more prolific and destructive, doing mischief by burrowing into the heartwood and leaving tunnels as large or larger than a pencil. Flatheaded borers damage or kill trees by mining beneath the bark or by tunneling into the heartwood and sapwood, leaving dead areas of bark with sap exuding.

Woodpeckers, especially the Yellowbellied Sapsuckers, cause damage to trees that is often attributed to borers. The bird will likely make a row of holes in a more-or-less straight line pattern circling the limb or trunk. The sapsucker revisits the tree many times, feeding on the sap accumulated in the holes it has drilled.

Roundheaded Borers = Order: Coleoptera; Family: Cerambycidae
Flatheaded Borers = Order: Coleoptera; Family: Buprestidae

_____ CONTROL CLUE _____

I have bad news again! Once borers have entered a tree, control is usually fruitless. Nothing can be applied to the soil, or sprayed on or injected into a large tree that is effective. If only a tree or two is involved, you might try digging them out with a stiff wire, but don't bet on it. A preventive spray program using Lindane or Dursban is really the only way to go.

Remember, shade tree borers attack only weakened or otherwise stressed trees, but a borer infestation is not necessarily fatal. Some trees, live oak for instance, acquire a resident population of borers and tolerate them for years with no apparent harm to the host.

Metallic Wood Boring Beetle adult
(VWR)

Flatheaded Borer
(DOF)

Longhorned Beetle; common in pine
(DOF)

A shorter horn, Longhorned Beetle
(DOF)

Roundheaded Borer
(DOF)

Emergence and entrance holes of a
Roundheaded Borer
(DOF)

Roundheaded Borer damage to oak
tree; note sap staining on bark.
(DOF)

Southern Pine Sawyer adult
(DOF)

SHOTHOLE BORERS

Shothole Borers are sometimes known as Fruit Tree Bark Beetles. An interesting "equation" might look like this: bark beetle = roundheaded borer = shothole borer = peach tree borer = fruit tree bark beetle = peach twig borer = American plum borer = boring insects. (Don't try to prove Einstein's Theory with *this* equation.)

The point is this: several boring insects exist that will attack stone-fruit and other trees. Shothole borers are beetles. Peach tree borers (not wood borers in the true sense), peach twig borers and American plum borers are small, clear-winged moths whose larvae are caterpillars which bore into the wood and do comparable dirty deeds. Inconsistent genealogy notwithstanding, the result is dead fruit tree buds, twigs and limbs and sometimes even dead fruit trees.

Female shothole borer beetles bore into the tree wood, creating tunnels in which they lay their eggs. The grubs that hatch bore into the inner wood, creating sawdust-filled burrows 2 to 4 inches long. The grubs pupate just under the bark, then emerge as adult beetles. The last generation of grubs spends the winter in the tunnels, emerging the following spring. Weakened, diseased, and dying trees and branches are most susceptible to borer infestation.

Shothole Borers = Order: Coleoptera; Family: Scolytidae

Shothole Borer evidence on bark
(USDA)

SQUASH VINE BORER

The Squash Vine Borer is my personal Garden Enemy #1. (It's a cousin to the peach tree borer.) If I had my way, I'd hang this crook's picture in the Post Office. This bug does not fight fair. Look, I've just begun to write about it and already I have a headache. Excuse me, while I go get an aspirin.

Thanks. I believe I can make it now.

The adult moth is metallic green-black colored with hind wings fringed with black and orange hairs. It has similar colored markings over much of the abdomen. It is a day flier. Larvae (worms) are white, heavy-bodied, and over an inch long when full grown. And it's the larvae that do the dirty work. Here's their act: Moths emerge in early summer and lay eggs on plant stems, normally close to ground level. On hatching, larvae bore into vines and that's the problem. That magnificent squash plant will wilt and soon die as the larvae tunnel up the hollow stem.

Squash Vine Borer = Order: Lepidoptera; Family: Sesiidae

CONTROL CLUE

Go after the adults with Sevin dust; begin treating as soon as you have a vine to treat. Keep the base of the plant well dusted. Pyrethrin spray will knock 'em down also, but there is no residual control. If you notice the wilt soon enough, find where the worm has entered the vine (it will be near ground level and will be covered with a wet, tan sawdust-like mess). With a sharp knife, slit the stem vertically till you find the worm. Dig him out. Pack soil around the wound and it will likely heal. Remember, "an ounce of prevention . . ." or you also will have a headache — and no squash.

Oh, me! I need another aspirin.

 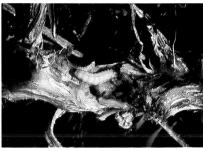

Squash Vine Borer adult
(TAEX)

Squash Vine Borer larva and damage
(TAEX)

_____Boxelder Bug_____

Boxelder Bugs and other closely related insects are found throughout most of Florida and they feed on several kinds of trees. In the fall they enter houses to find shelter for the winter and thus become a nuisance to many homeowners. They are plant feeders (flowers and leaves) and do not feed on structures, food products, cloth, humans or pet animals; however, these bugs may stain curtains, paper and similar furnishings with fecal spots. They produce little or no odor when crushed.

Boxelder bugs are about ½ inch long and ⅓ inch wide. They are brownish-gray to black with distinctive red markings. (Check the bright red abdomen.) They commonly hide in cracks and crevices in walls, door and window casings, and around foundations, as well as in tree holes and in large accumulations of debris. Adults are likely to come out of hiding and appear on light or white painted surfaces during warm days of winter or early spring. Adults deposit eggs in cracks and crevices of tree bark in the spring about the time buds begin to open.

Another insect, Jadera haematoloma (no common name), is often confused with the boxelder bug because it is so similar in appearance and habits. This "Jh" is a beautiful bluish to smoky-black color with bright red eyes and stripes along the head. Apparently the development of this critter is similar to that of the boxelder bug.

Boxelder Bug = Order: Hemiptera; Family: Rhopalidae

_____ CONTROL CLUE _____

Boxelder Bugs are not considered to be serious pests to host plants. Although they feed by sucking plant juices, they are seldom abundant enough to harm trees. If insecticide treatment is necessary, it should be applied to young, exposed bugs found on host trees in the spring or early summer. Use an approved Dursban or Diazinon spray. Read the label.

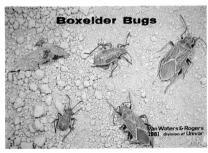

Boxelder Bug adult
(VWR)

Boxelder Bug eggs on boxelder seed, nymphs and adults
(VWR)

―――――Cabbage Looper―――――――――

Cabbage Loopers, one of the worst of all loopers, are voracious feeders. They can strip foliage from an infested plant in short order. They enjoy a diversified menu, like cabbage, cauliflower, broccoli, Brussels sprouts, lettuce, and occasionally beans, tomatoes and other crops. Larvae (worms) are light-green caterpillars with a few white or pale yellow stripes. They travel with a characteristic looping motion, hence the name . . . Looper. Watching a looper travel tends to remind you of the hiccups.

Cooler temperatures slow their activity as it does with all other insects. There are continuous generations in southern Florida where typically temperatures are uniformly tepid. Good ol' Mother Nature hits a lick for our side in that, when populations become crowded, a virus disease often strikes causing high larval mortality. Otherwise, I guess loopers would rule the world.

Cabbage Looper = Order: Lepidoptera; Family: Noctuidae

―――――― **CONTROL CLUE** ―――――――――

Sevin, Diazinon or the bacterium, *Bacillus thuringiensis*. (Don't worry with remembering *Bacillus thuringiensis*. Ask for Dipel or Thuricide . . . same thing.)

Cabbage Looper larva
(USDA)

Imported Cabbageworm

The Imported Cabbageworm's attack is similar to and easily confused with that of its kissin' cousin, the cabbage looper. Larvae are dark green caterpillars growing up to 2 inches long. They are frequently found right along with the looper, but there is one very obvious difference between the two. The imported cabbageworm travels without that looping motion . . . no hiccups. The menu and appetite of the two are about the same. They eat chosen vegetable leaves like crazy.

There is another most distasteful characteristic of cabbage caterpillars. Their toilet habits are atrocious. Another form of damage by them is contamination of plants with greenish-brown excrement. And the bigger the worm, the bigger the . . . well, you know what!

Imported Cabbageworm = Order: Lepidoptera; Family: Pieridae

CONTROL CLUE

Like for the cabbage looper: Sevin, Diazinon, Dipel or Thuricide. Read the label.

Imported Cabbageworm larva
(USDA)

Carpenterworm

Carpenterworms bore into the wood of living hardwood trees. The large winding tunnels in the sapwood and heartwood constructed by the larvae serve as an entrance for wood-rotting fungi and insects such as the carpenter ant. In cases of extreme infestation, the tree may be structurally weakened and subject to wind breakage. In Florida, oak trees and carpenterworms are like soup and sandwich . . . they just go together. Oak species are preferred hosts, but cottonwoods, maples, willows, other hardwoods and sometimes fruit trees are also attacked.

Adult moths emerge in late spring to early summer, mating occurs, and then the females lay groups of eggs in bark crevices or wounds. Each mama lays 200 to 500 eggs during her one-week life span. After hatching, the larvae wander over the bark for a short time and then bore into the inner bark where they feed until they are half grown. They then bore into the sapwood and heartwood, returning occasionally to feed in the inner bark.

The larval period lasts from two to four years. Pupation usually occurs deep within the heartwood. Just prior to emergence, the pupa wiggles to the entrance hole where it remains slightly protruding and plays peek-a-boo until the adult moth emerges.

If you seek early symptoms, what you might first notice will be a sap-stained hole or holes in the bark that could have been made by a 30-06 rifle slug; you might also notice wood chips and pellets at the base of the tree.

Carpenterworms = Order: Lepidoptera; Family: Cossidae

CONTROL CLUE

If you can catch this sucker in the larval stage when it's wandering over the bark, you've got a chance for effective control. Try a good soak of Lindane or Dursban on the tree's bark. Once it bores into the sapwood and heartwood, better you spend your time on a formula for retiring the national debt. Some chemicals which have a fumigating action have proved somewhat effective in controlling this insect in shade trees. No practical control has yet been found for forest trees. If you find an acceptable control, send it to me.

Carpenterworm moths, male and female
(SFI)

Carpenterworm larva in oak
(SFI)

Carpenterworm gallery in oak made by two year old larva
(SFI)

Carpenterworm entrance hole in oak
(DOF)

Caterpillars

CABBAGE PALM CATERPILLAR

The Cabbage Palm Caterpillar, sometimes referred to as the Cabbage Palm Worm, is the larva of an owlet moth. The larvae sometimes gather by the thousands on immature palm buds and completely denude the trees of all trace of blooms as they feed. When heavy infestations occur, mature caterpillars have been known to invade homes and destroy

69

household fabrics. In making their cocoons, they utilize any available fabric, such as rugs, draperies, stuffed furniture, bedding and clothing. An outbreak in Miami, Florida caused serious loss when caterpillars began using fiberglass cloth used by the boat-building industry.

The mature larva is about 1½ inches long; skin is covered with very small black spines which are not visible to the naked eye. The body appears pinkish in color and has a shiny black head. The larva becomes very pink just before pupating. The dark reddish-brown pupa is protected by a tough cocoon. The adult moth is fawn colored with a wing expanse of about 2 inches; a dark eyespot about 3/16 inch in diameter is found on each hind wing.

High populations of the Cabbage Palm Caterpillar can drastically reduce the amount of palmetto honey produced in Florida. Tiny larvae may be found deep in developing bloom spikes. Fully mature caterpillars drop to the ground on silken threads and crawl to protected pupation sites, or may crawl down the trunk to the base of dead fronds where they build cocoons from dry palm fibers. Back in 1960, near Geneva, Florida, palms put forth a heavy bloom and a late larval population completely wiped out all visible blooms.

Cabbage Palm Caterpillar = Order: Lepidoptera; Family: Noctuidae

CONTROL CLUE

A 50% Sevin WP spraying of your palm trees should do the job, but Sevin is really rough on the bee population, so maybe Dipel or Thuricide would be better. Whatever you do, do it early; remember, little worms are easier to kill than big worms. Those big larvae sound like good big-mouth bass bait to me. Are you thinkin' what I'm thinkin'?

Cabbage Palm Caterpillar — top view
(DOF)

SALTMARSH CATERPILLAR

We do have our share of caterpillars in Florida. And here's another that you hear a lot about. Saltmarsh Caterpillars eat everything — most garden crops, field crops, weeds, shrubs; they're sure not particular.

Adults are white moths with black-freckled wings. Mama's wings are yellow on the underside. Papas have hind wings which are yellow above and below. These dudes are prolific breeders and make a lot of babies (worms). Larvae are up to 2 inches long when full-grown and are covered with dense hairs ranging from yellowish to brown to nearly black in color. They migrate in a manner similar to armyworms, stripping foliage like a lawn mower. Saltmarsh caterpillars can boogie through your garden before you bat an eye.

Saltmarsh Caterpillar = Order: Lepidoptera; Family: Arctiidae

_____ CONTROL CLUE _____

A shot of Diazinon spray thoroughly applied should do the job.

Mating parents of Saltmarsh
Caterpillars (Acraea Moths)
(VWR)

Saltmarsh Caterpillar larvae —
"wooly-worms"
(TAEX)

Saltmarsh Caterpillar 1st instar larvae
on sycamore leaf
(VWR)

Saltmarsh Caterpillar on pine seedling
(DOF)

STINGING CATERPILLARS

Several kinds of stinging caterpillars occur in Florida, but the most common venomous larva is the fuzzy, white, tan or gray PUSS CATERPILLAR, which is often incorrectly called an "asp." The IO MOTH larva, the SADDLEBACK CATERPILLAR and the HAG MOTH larva can hurt you also.

These caterpillars are the immature stages of various moths whose venom is conveyed through stiff, hollow, hair-like spines. These critters feed upon the foliage of many plants, including trees, shrubs and flowers. They rarely occur in numbers sufficient to damage plants, but they are important because of their medical effects.

A person's first symptom following contact with a puss caterpillar is an intense burning at the site of contact. The severity of the symptoms appears to be related to individual sensitivity, maturity of the larva, number of spines contacted, degree of pressure against the caterpillar, and site of the sting. In sensitive persons, lymph nodes under the arms or in the groin may swell and hurt; nausea, weakness and shock-like symptoms with severe headache may occur, usually within two hours after contact. Don't hesitate to see your physician. Prompt application of an ice pack and a baking soda poultice may help to reduce pain and prevent swelling.

Don't rub the affected area. Use the sticky side of Scotch tape to lift the spines off your skin; Adolph's Meat Tenderizer applied to the stung area might offer some relief. There is no really effective home first-aid treatment, but at least you will have tried.

Puss Caterpillar = Order: Lepidoptera; Family: Megalopygidae
Io Moth = Order: Lepidoptera; Family: Saturniidae
Hag Moth = Order: Lepidoptera; Family: Limacodidae
Saddleback Caterpillar = Order: Lepidoptera; Family: Limacodidae

---

CONTROL CLUE

Where stinging caterpillars present hazards to persons, such as around schools or residences, infested shrubs and trees may be sprayed or dusted with Dipel, Thuricide or Sevin. Remember, don't touch the critter.

Puss Caterpillar close up
(TAEX)

Puss Caterpillar (tan)
(TFS)

Puss Caterpillar. Can you find it?
(DPI)

Saddleback caterpillar
(DPI)

Hag Moth Caterpillar
(DOF)

Io Caterpillar
(DOF)

TENT CATERPILLARS

EASTERN TENT CATERPILLARS build large webs; FOREST TENT CATERPILLARS build no tents at all. (I think they have labor problems!) Both caterpillars are closely related and are particularly troublesome in Florida.

Tent caterpillars feed in groups and can defoliate a tree in short order. Mamas are moths. Larvae hatch in early spring at about the time host plants leaf out. Tents of the Eastern are most often found in a crotch of small limbs at the trunk of the tree and these tents enlarge as the colony grows. Larvae move from these refuges to feed on leaves, so damage can be found for some distance from the web.

Tent caterpillars are rather like people in the sense they all do about the same things, but differ primarily in appearance. Larvae (worms) are attractively colored, are about 1½ inches long, and have a few long white hairs on their bodies, mostly along the sides. The Eastern has a solid white line down the center of its back; the Forest has a row of "key-hole" shaped white marks, one on each body part.

Common host trees of the eastern tent caterpillar in Florida are apple, black cherry, hawthorn, oaks, pecan and plums; the forest tent caterpillar has the "hots" for black tupelo, oaks, sweetgum, water tupelo, plums and other hardwoods.

Eastern Tent Caterpillar = Order: Lepidoptera; Family: Lasiocampdiae
Forest Tent Caterpillar = Order: Lepidoptera; Family: Lasiocampdiae

CONTROL CLUE

There is only one generation of tent caterpillars a year. If they have been allowed to feed and have completed their development, it is useless to spray, so hit 'em early with Sevin, Thuricide, Dipel, Diazinon, Malathion or Methoxychlor. There's plenty of ammunition out there for these critters, so load up and shoot. But do it carefully.

Eastern Tent Caterpillars — solid line
Forest Tent Caterpillars — foot prints
(TFS)

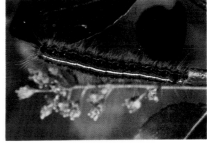

Eastern Tent Caterpillar larva
(USDA)

Forest Tent Caterpillar after shower
and shave
(TFS)

Forest Tent Caterpillar adult moth on
oak leaf
(VWR)

75

Tent mass of Eastern Tent
Caterpillar
(SFI)

Forest Tent Caterpillar horde headin'
for home . . . and lunch
(DOF)

VARIABLE OAK LEAF CATERPILLAR

The Variable Oak Leaf Caterpillar periodically defoliates hardwood trees in Florida. The larvae feed primarily on oaks, but will also feed on apple, black walnut, persimmon, hawthorn, birch and elm. Young larvae skeletonize the leaf while older larvae devour the entire leaf except for the primary veins. While infestations usually subside before many trees are killed, heavy defoliation reduces the tree's vigor and growth.

The variable oak leaf caterpillar overwinters as a non-feeding larva in a cocoon on the ground. It pupates and emerges as a moth the following spring. The female moth, gray in color and about 1¾ inches long, lays about 500 eggs singly on the leaves of host trees. The hatched larvae feed on foliage for five or six weeks, drop to the ground to pupate and emerge as adults in mid-summer. Larvae hatching from eggs laid by the second generation of moths defoliate the trees for the second time during late summer. By late October, the mature larvae of the second generation have dropped to the ground to overwinter.

The full-grown larva is approximately 1½ inches long. The body is usually yellow-green with a narrow white stripe down the center of the back bordered by wider dark bands which often are a hot pink. (Is this beginning to sound like a designer fashion show?) The head is usually amber-brown with curved diagonal white and black bands. Colors will vary among individuals.

_____ CONTROL CLUE _____

Outbreaks of the variable oak leaf caterpillar may be severe, but usually subside before tree death occurs. Although mice and predaceous beetles feed on the resting larvae and pupae, other predators and parasites generally are not effective in controlling rising populations of this critter. Sevin, Thuricide or Methoxychlor have been effective and safe in controlling related caterpillars.

Caution: Variable oak leaf caterpillars produce formic acid when disturbed. Excessive handling may result in skin irritations for susceptible people.

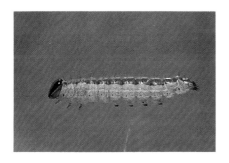

Variable Oak Leaf Caterpillar
(DOF)

Variable Oak Leaf Caterpillar adult
(SFI)

Variable Oak Leaf Caterpillar prepupae
in soil, getting ready for winter
(DOF)

Variable Oak Leaf Caterpillar larvae
and damage
(SFI)

WALNUT CATERPILLAR

Walnut Caterpillars can be a serious threat to pecan, hickory and walnut trees. Oak, willow, honey locust and certain woody shrubs can be endangered also. Two generations usually develop each year, with the second being more prolific, thereby causing more damage. A severe outbreak can cause devastating defoliation of host trees in a given area. Walnut caterpillars can be bad news, but fortunately they don't usually hit Florida flora that severely.

The adult is a moth with a 2-inch wing span. It is brown and tan with a dark region on the body behind the head with wavy, dark lines across the front wings. The female will deposit about 300 eggs on the underside of a leaf. Caterpillars (larvae) hatch in about nine days and live together in a group, growing up to 2 inches in length. They do not make webs. They are reddish-brown to black with white markings and have long, fuzzy white hairs. Larvae characteristically arch their heads and tails in a defensive posture when disturbed. When messed with, they "U" at you.

Walnut Caterpillar = Order: Lepidoptera; Family: Notodontidae

_____ CONTROL CLUE _____

To avoid as much damage as possible, spray these caterpillars when they are still young and small. Thoroughly wet leaves, twigs, limbs and tree trunks with Sevin, Orthene, Diazinon or Thuricide, the choice depending upon variety of tree. Remember, read the label on the container and follow directions. Large trees may require high-pressure spraying for effective control. Don't waste any time if you find them preparing to picnic on one of your trees.

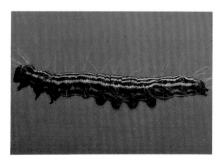

Walnut Caterpillar after shave and haircut
(DOF)

Disturbed Walnut Caterpillars; note
arched heads and tails.
(TFS)

Walnut Caterpillar orgy on a pecan tree
(TAEX)

YELLOWNECKED CATERPILLAR

The Yellownecked Caterpillar is a tree defoliator; oaks and hickories are favorite hosts, but apple, elm, maple, pecan, walnut and other hardwoods are also attacked. Adults emerge from the soil by late spring, mate, and then the females lay eggs on the undersides of leaves. When larvae hatch, they feed together in colonies during the summer and fall. There is only one generation per year.

Larvae are readily identified by distinctive markings — black head, bright orange to yellow neck, black body with 8 thin yellow to white stripes and sparse, long, white or gray hairs. Biology, habits and signs of infestation are very much like those of the walnut caterpillar; they also rear into a U-shape when disturbed. The major difference is, of course, in appearance.

Yellownecked Caterpillar = Order: Lepidoptera; Family: Notodontidae

_____ **CONTROL CLUE** _____

Orthene will do them in. And don't let them eat a bunch before you spray; the more they eat — the more they damage.

79

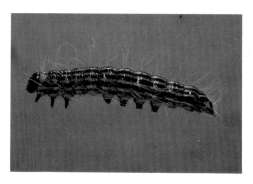

Yellownecked Caterpillar with "ring around the collar — ring around the collar"
(DOF)

Catfacing Insects

Catfacing is a disfiguration of fruit mostly caused by a group of sucking insects and some chewing insects. It is a major problem in Florida. This malformation is caused by these "bandidos" attacking the fruit when it is in an early, formative stage. These insects, which are actually various stink bugs and plant bugs, penetrate the fruit with their sucking mouthparts and cause this scarring which we call "catfacing."

The TARNISHED PLANT BUG is a major pest in Florida peach and plum orchards, but it will also invade your yard to do its mischief. The adult is brown and mottled with yellow, black and red. It is about ¼ inch long. Nymphs are green.

Early in the season, at the pink bud stage, LYGUS BUGS commonly attack peaches. These insects are about 1/5 inch long and range in color from shades of brown to tan to nearly black. They damage buds and blossoms as well as small fruit.

In addition, several species of STINK BUGS attack peaches and plums in Florida. Their damage may occur early in the fruit's development, but the blemishes remain on the final product. Damage caused by the feeding of the PLUM CURCULIO can also cause catfacing in most fruits, particularly in peaches, plums and apples.

Hail or cold weather damage to tender blooms may also cause catfacing. Tomato catfacing can be caused by extreme heat (above 85 degrees Fahrenheit) or by cold (below 55 degrees Fahrenheit) or by drought.

Tarnished Plant Bug = Order: Hemiptera; Family: Miridae
Lygus Bug = Order: Hemiptera; Family: Miridae
Mirid Plant Bug = Order: Hemiptera; Family: Miridae
Stinkbugs = Order: Hemiptera; Family: Pentatomidae

CONTROL CLUE

Mother Nature will determine the temperatures, but we can handle those insects. Spray with Malathion 50% E.C. (emulsified concentrate-liquid) when ¾ of the bloom petals have fallen. Clean up weeds and plant debris in the fall to eliminate hibernation locations for overwintering bugs.

Harlequin and Green Stinkbugs
(VWR)

Tarnished Plant Bug adult
(VWR)

Lygus Bug adult
(VWR)

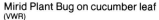

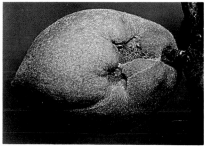

Mirid Plant Bug on cucumber leaf
(VWR)

Catfacing plant bug injury on peach
(USDA)

_____Centipedes and Millipedes_____

Centipedes and Millipedes are usually considered nuisances rather than destructive pests. Centipedes pose an occasional threat to man because they are equipped with poison glands and they will bite. The poison usually produces only a moderate reaction similar to that of a bee sting, but if you are bitten and have a significant reaction, don't hesitate to see your physician. Most centipedes feed upon small creatures such as insects. With powerful jaws, which are located immediately behind their head, they grasp their prey and kill it by injecting venom. They prefer moist, protected habitats such as under stones, rotted logs, leaves, bark or in compost piles.

Millipedes are not poisonous, but many species have repugnatorial glands capable of producing mild acids which may produce allergenic reactions in sensitive individuals. A few millipede species are capable of squirting these fluids over a distance of several inches. Persons handling millipedes will notice a lingering odor on their hands and the fluid can be dangerous to the eyes, so don't fool around with them. Millipedes feed primarily on decaying organic matter, and rarely do they attack roots and leaves of seedling plants, but the experts still argue about this. Their favorite abode is the greenhouse. The two species most frequently encountered in Florida are the Hothouse or Greenhouse Millipede and the Tropical Millipede.

Centipedes = Class: Chilopoda; Order: Scutigeromorpha; Family: Scutigeridae
Millipedes = Class: Diplopoda; Order: Chordeumida; Family: Lysiopetalidae

---

CONTROL CLUE

Contact sprays of Pyrethrins may be applied directly to centipedes and millipedes for quick control inside the home — or even quicker, hit 'em with a shoe. Sevin or Diazinon granules or baits containing Mesurol do well outdoors.

Centipede
(VWR)

Millipedes: one walking; one coiled
(VWR)

Chiggers

Common Chiggers, also known as "redbugs," attach themselves to tender skin areas to feed. These very tiny pests belie their size in the amount of torment they cause humans. Their bites cause intense itching and small reddish welts on the skin. Most persons cannot see chiggers without a magnifying glass, so the bites may be the only indication that the bugs have infested a given area.

Chiggers are related to ticks and spiders. The young chigger (larva) which hatches in the spring is the real troublemaker, and it can wait a long time for just the right meal to show up — namely you. It is a parasite that feeds on man and animals. This wee, bright red larva can scarcely be seen as it scurries along the skin surface seeking an attachment site. When it finds one, such as a skin pore or hair follicle, it attaches its mouthparts to the spot. On people, the chigger prefers places where clothing fits tightly or where the flesh is thin or wrinkled. (Anybody over 10 years of age has some wrinkled skin somewhere.)

Contrary to common belief, it does not burrow into the skin or suck blood. Instead, it injects a digestive juice that disintegrates skin cells so they can be used as food. Affected skin tissue becomes red and swollen. It may completely envelop the feeding chigger, making it appear to be burrowing into the skin. The bite itches intensely and may continue itching for several days after the chigger is killed or drops off.

Chigger = Class: Arachnida; Order: Acari; Family: Trombiculidae

---

CONTROL CLUE

Mosquito repellents are usually effective against chiggers also. Common dusting sulphur, although somewhat messy and smelly, is also a fairly effective repellent. After exposure to chigger-infested areas, take a hot soapy bath to kill and remove larvae; then apply an antiseptic solution to any welts that might appear. Destroying the chigger normally does not stop the itching, but it will cease usually after two to three days. Temporary relief can be obtained by applying a mild topical anesthetic. Your druggist can suggest an appropriate product. If you are forced to a chemical control for your lawn, go with a Dursban or Diazinon spray.

Chigger — not magnified (Don't gripe! I told you they were little.)

_____Chinch Bug, Southern_____

The Southern Chinch Bug is the most injurious and notorious pest of St. Augustine grass in Florida. In northern Florida, eggs begin hatching about mid-April each year marking the beginning of 3 to 4 generations. In south Florida, eggs begin hatching in late February, with 7 generations per year. The nymphs (babies) are about the size of a pinhead when they hatch. They are yellow and wingless, but soon turn red and develop a light-colored band across their backs. Before the last molt, nymphs are black or brownish-black with a white spot and two small wing pads on their backs. Adults are about 1/6 to 1/5 inch long, with black bodies and white wings. Each wing bears a triangular black mark.

Southern chinch bugs suck the plant juices from grass, resulting in yellowish to brownish patches in infested lawns. These injured areas are often first noticed in water-stressed areas along edges of lawns, especially during dry periods. Other factors such as disease, nutritional deficiencies and drought will also cause these symptoms; therefore, chinch bug infestations can be accurately diagnosed *only* if the insects can be found.

So, how do we find chinch bugs? When damage is severe and bugs are plentiful, they usually can be found simply by parting the grass and carefully observing the soil surface. Another way chinch bugs can be found is by pressing one end of an open-ended coffee can about 2 inches into the soil at the edge of the beginning-to-yellow grass. Fill the can with water and keep it nearly full for about 5 minutes. Any chinch bugs present will float to the surface and do a six-legged breast stroke that would make an Olympic swimmer green with envy.

Southern Chinch Bug = Order: Hemiptera; Family: Lygaeidae

_____ CONTROL CLUE _____

Several predatory insects are known to attack chinch bugs. The most prominent is a black Bigeyed Bug. A predacious earwig (Labidura) is also a very good predator on all stages of the chinch bug. If you find a bunch of these good-guys, hold off using an insecticide and let them do your dirty work. When the damage threshold is reached (20 to 25 chinch bugs per square foot) a pesticide should be applied. Dursban or Diazinon is good, but in several areas of southern Florida chinch bug populations have apparently developed a resistance to these insecticides, so you might have to go with Baygon or Oftanol.

Southern Chinch Bug adult
(DOW)

Southern Chinch Bug red nymph
(DOW)

Chinch Bugs, males and females
(VWR)

Chinch Bug adult and nymph on grass
(VWR)

_____Cicadas_____

Cicadas are frequently misidentified as locusts. Even today their appearance arouses fear of crop destruction. American Indians once thought these curious creatures had an evil significance. These critters are more frequently heard than seen, being responsible for that screeching cacophony during hot, late-summer afternoons. Only the males sing . . . if you can call that noise singing . . . and they do it for the purpose of attracting females to mate.

The females' egg-laying activity can severely damage or destroy twigs and small branches of vines, shrubs and trees. Eggs hatch in 6 to 7 weeks. The resulting ant-like nymphs drop to the ground and burrow through the soil to find roots from which they suck juices. There are 7-year cicadas, 13-year cicadas and 17-year cicadas. The nymphs will stay in the ground that long, just suckin' juices, before they finally emerge as adults. Man, talk about patience!

Under some conditions, the nymphs may build mud cones or chimneys 3 or 4 inches high from which they emerge from underground. The emergence hole in the mud cone (or in the soil if no cone is constructed), is approximately ½ inch in diameter. Adult emergence from the nymphal skin takes place at night on trees, posts or buildings. This light brown, brittle, beetle-shaped shell that you are so reluctant to touch will hang on a tree trunk for months if left undisturbed. It's harmless.

Cicadas = Order: Homoptera; Family: Cicadidae

```
┌─────────── CONTROL CLUE ─────────────────────────────┐
│                                                      │
│   General chemical control of the cicada normally is not │
│   justified. If you have a special plant that deserves protection │
│   from that bark-splitting, egg-laying mama, apply Sevin on a │
│   weekly schedule. Not much you can do about that incessant │
│   screeching. Write it off as a sound of summer and find │
│   something else to worry about.                     │
│                                                      │
└──────────────────────────────────────────────────────┘
```

Cicada egg-laying damage
(DOF)

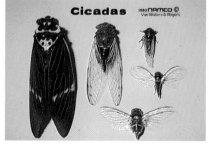

Five different species of Cicadas; size range — ½ to 3 inches
(VWR)

Cicada adult
(DOF)

Cicada nymphs
(DOF)

Cockroaches

Cockroaches are the critters people love to hate. Approximately 3500 species exist in the world today, and Florida seems to have more than its share. The next time you hear someone complaining about the number of cockroaches in Florida, remind that critic: we didn't choose them . . . they chose us! I guess that shows cockroaches have good taste.

Cockroaches, also known as waterbugs, croton bugs or palmetto bugs, destroy food and damage fabrics, bookbindings and other materials. They secrete an oily liquid from scent glands that is offensive; this odor may ruin food and be imparted to dishes that are apparently clean. Leaving excrement in the form of pellets or an ink-like liquid is another of their distasteful habits. Cockroaches are a bane of Florida orchid growers and can cause considerable damage to other greenhouse plants and flowers by chewing them.

One of the most common native palms in Florida is the saw palmetto, a low-lying (groundrunner) palm. These and other palms provide the perfect haven for cockroaches, i.e., darkness, moisture, plenty of decaying organic matter and protection from predators. In olden days, there was a profusion of saw palmetto around Florida Cracker homes and these plants were infested with "palmetto bugs" which were most likely Florida Woods or American cockroaches. During cooler weather, or just because, these bugs would move into a house. The residents put two and two together and linked these circumstances forevermore, by calling the cockroaches — PALMETTO BUGS.

Cockroaches prefer a moist environment; consequently, they will hold a class reunion in a pile of wet wood, a sewer pipe or any location that is wet . . . so calling them WATER BUGS comes easy. And Water Bug just sounds much less threatening than Roach!

The cockroaches most commonly found in and around Florida homes are: FLORIDA WOODS, AMERICAN, SMOKY BROWN, BROWN, AUSTRALIAN, GERMAN AND BROWN-BANDED. (This team could win the Super Bowl.) The smallest of this group are the German and the Brown-Banded; the adults are seldom more than ⅝ inch long. The larger cockroaches, the American, Australian, Brown and the Smoky Brown, are 1½ to 2 inches long. These are generally found outdoors, but can become an indoor problem when they migrate or are brought indoors. The largest of this group, the Florida Woods, will also enter dwellings from outdoors.

In addition to the group of cockroaches that commonly call Florida "home," you might also encounter some of their kinfolks from out of town: ORIENTAL, CUBAN, PENNSYLVANIA WOOD, LOBSTER or CINEREOUS, SURINAM and CUBAN DEATH'S HEAD.

All cockroaches are nocturnal; they hide in dark, damp places during the day and search for food at night. If disturbed, they run rapidly for shelter. Cockroaches are some of the oldest insects, as indicated by fossil remains dated to 200,000,000 years ago. We contend with a formidable adversary in attempting to eliminate this pest, but there's no choice — either fight him or give him a deed to the house.

German Cockroach = Order: Dictyopera; Family: Blattellidae
Brown-banded Cockroach = Order: Dictyopera; Family: Blattellidae
Oriental Cockroach = Order: Dictyopera; Family: Blattidae
American Cockroach = Order: Dictyopera; Family: Blattidae
Australian Cockroach = Order: Dictyopera; Family: Blattidae
Brown Cockroach = Order: Dictyopera; Family: Blattidae
Smoky Brown Cockroach = Order: Dictyopera; Family: Blattidae
Florida Woods Cockroach = Order: Dictyopera; Family: Blattidae
Lobster Cockroach = Order: Dictyopera; Family: Blaberidae
Surinam Cockroach = Order: Dictyopera; Family: Blaberidae
Cuban Cockroach = Order: Dictyopera; Family: Blaberidae
Cuban Death's Head Cockroach = Order: Dictyopera; Family: Blaberidae

Study each cockroach problem and use control measures in accordance with the location, extent and nature of the infestation. Hit 'em with everything available, including your shoe; put out roach baits; use contact insecticides and residual sprays. I remember the sage advice of my friend, Bill Spitz: "S.O.S. — Seek Out the Source." Employing a capable professional exterminator will be your easiest solution.

If you want a surefire check for roaches in your home, invite the snobbiest people you know over for dinner. If there's a roach in the house, I promise it will run across the floor that evening. And if you see a roach, don't scream R-O-A-C-H. Quietly say, "My, what is that?" Your husband should reply, "Oh, just a water bug." Don't lose your cool!

American Cockroach
(USDA)

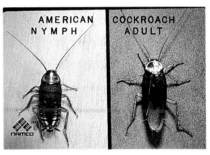

American Cockroach nymph and adult
(VWR)

American Cockroach and house mouse
droppings comparison
(VWR)

German Cockroach
(USDA)

90

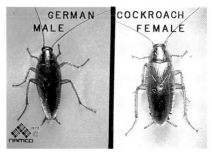

German Cockroach adult male/female comparison
(VWR)

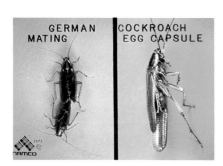

German Cockroaches mating and female with egg capsule
(VWR)

Smoky Brown Cockroach
(VWR)

Brown-Banded Cockroach
(USDA)

Cuban Cockroach
(IFAS)

Oriental Cockroach
(IFAS)

Madagascarn Cockroach
(DOF)

Australian Cockroach
(IFAS)

Surinam Cockroach
(DOF)

Lobster or Cinereous Cockroach
(DOF)

Top from left: Florida Woods,
American, Smoky Brown, Brown and
Australian
Bottom from left: German and
Brown-banded
(IFAS)

Comperia merceti predator stinging
roach egg capsule
(VWR)

92

Death's Head Cockroach (female)
(DOF)

Death's Head Cockroach (male)
(DOF)

_____Conenose, Blood–Sucking_____

The Blood-sucking Conenose occasionally enters homes and sucks blood from sleeping persons, and for this reason it is often called the "big bed bug" or another colloquial name. The feeding bites are usually benign; however, individuals who have an allergic sensitivity may experience dizziness, nausea, burning pain, intense itching and much swelling along with red blotches and welts over the body. At the least, you'll think about calling your doctor for some lab work. Effects from a bite may last for months, but usually disappear after a few days.

The blood-sucking conenose is most dangerous because of its ability to transmit the sometimes fatal Chagas disease which primarily affects children. The most apparent symptoms are fever and swelling of the eyelids and face.

In domestic situations, the blood-sucking conenose may be found in bedding, in floor cracks and walls, under furniture, in hen houses, outhouses, barns and doghouses. Natural habitats include hollow trees and stumps, on palmetto boots and trunks, under bark of oak and pine and in wood rat nests.

Blood-sucking Conenose = Order: Hemiptera; Family: Reduviidae

93

The adults are night fliers and are attracted by lights, so be sure open windows are adequately screened. Eliminate breeding places. Apply an approved aerosol spray in rooms and follow up with a residual spray to all potential hiding places. If one of these critters alights on you, brush him off quickly. Do not take hold of him because the possible resulting stress bite will be painful. This monster and his lifestyle would be a great subject for a late-night TV horror movie.

Blood Sucking Conenose adult
(IFAS)

Blood Sucking Conenose nymph
(IFAS)

_____Coneworm, Southern Pine_____

Southern Pine Coneworm adults, which are moths, can cause severe losses in seed orchards with the destruction of flowers, conelets and cones. Death is uncommon, but saplings with a high attack rate may die from the girdling caused by larvae feeding in the inner bark. Adults become active in the early spring, mate, and then the females lay eggs on or near cones, shoots or wounds. There may be 1 to 4 generations per year. Common hosts include bald cypress, loblolly pine, longleaf pine, pond cypress, sand pine, shortleaf pine, slash pine and spruce pine.

Larvae will vary in size, but coloring is a good identity clue; they usually have a dark head and the upper surface of the body will be a red-purplish brown with a greenish undersurface. The larvae do the dirty work, feeding within the cones or inner bark of shoots, branches or stems. There might be a tan pupal case protruding from the large pitch mass which is usually a milky white color with little, if any, redness or pinkness.

Although the Southern Pine Coneworm is of more economic concern in a pine seed orchard, it is not above attacking a favorite backyard pine tree or even a shade tree. And when you see that big blob of pitch exuding from your pine, you are likely to go nuts fighting a pine bark beetle that isn't there. Remember, coneworm adults are *moths*, not beetles. Pay attention.

Southern Pine Coneworm = Order: Lepidoptera; Family: Pyralidae

_____ **CONTROL CLUE** _____

Maintain a healthy tree. Avoid injury to the tree's bark; cone-worm moths will be attracted to a recent bark injury area. Prune branches that have pitch masses. A tree spray containing Lindane or Dursban will be worthwhile protection for a valued tree. If this problem "drives you up a tree," don't hesitate to call your local forester or county extension agent for help.

Southern Pine Coneworm adult moth
(SFI)

Southern Pine Coneworm pitch mass
(DOF)

Mature larva of Southern Pine
Coneworm feeding in slash pine cone
(SFI)

Southern Pine Coneworm
mature larva in loblolly pine
cone
(SFI)

Shoot damage by Southern Pine
Coneworm
(DOF)

_____Corn Earworm_____

The Corn Earworm, alias Tomato Fruitworm, alias Cotton Bollworm, alias Sorghum Headworm . . . all the same critter! The reason the corn earworm doesn't know he's a corn earworm is simply the fact that he doesn't read these insect books we humans write. This caterpillar is the most serious pest of corn, but also it is most devastating to tomatoes and often to cabbage, beans and peppers. This maniac is not above slipping over to your rosebush for yet another course on its menu.

Larvae (worms) may reach an inch and a half in length and may vary in color from green to pink to almost black. Alternating longitudinal dark and light stripes mark this worm's body, but colors are so variable that such characteristics are not dependable for identification. You will best know him by the fruit he eats. And can he eat. Remember, this caterpillar tunnels into fruit and/or feeds on leaves. Good luck with this one.

Corn Earworm = Order: Lepidoptera; Family: Noctuidae

_____ **CONTROL CLUE** _____

Sevin is about as good a control as any, but remember to hit him when he's little and damage is first noticed. Remember, no control . . . no tomatoes . . . no salad . . . and there goes the ol' summer diet.

Corn Earworm with corn damage
(USDA)

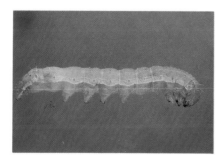

Corn Earworm
(DOF)

Sorghum Headworm
(USDA)

____Crickets_____

HOUSE & FIELD CRICKETS

Of the several species of crickets, House Crickets and Field Crickets are the most common and are the most troublesome. Both of these crickets have antennae that are longer than the body and both are good jumpers and fliers. Males have two appendages extending from the tip of the abdomen; females have three. Males make the noise.

Field Crickets prefer to live and breed outdoors where they feed on several kinds of plants. Occasionally, they invade homes in search of warm hiding places, but will not breed or establish a permanent infestation indoors.

House crickets commonly breed outdoors, but unlike field crickets, can live and breed indefinitely indoors. The fabled Jiminy Cricket notwithstanding, these rascals are pests. Outdoors, they damage garden plants. Indoors, they can damage woolens, cottons, silks, synthetic fabrics, furs and carpeting. Clothes stained with perspiration are particularly attractive. They also feed on foods, leather and rubber products.

Camel Crickets, also called Stone or Cave Crickets, are not true crickets, are wingless, and are otherwise distinctive in appearance because of their arched backs. They are active at night, are not attracted to light and have no "song."

Jerusalem Crickets are sometimes called "children of the earth." They have large, round, naked heads with two bead-like black eyes that may give them a fancied resemblance to a miniature child. They burrow into loose soil, are active at night and are seldom seen. They are useful predators, feeding on other insects and spiders. Although fierce looking, Jerusalem Crickets are harmless, but their powerful mandibles could inflict a minor, nonvenomous wound if handled carelessly.

Crickets = Order: Orthoptera; Family: Gryllidae

_____ CONTROL CLUE _____

INDOORS: Apply a residual spray of Dursban, or find the critter at night with a flashlight and bop him with a shoe, unless you are superstitious about killing crickets. (First-degree-cricket-murder is supposed to bring bad luck, but if you apply an insecticide for roaches and Brother Cricket happens to wander into it and dies, then it's either accidental death or suicide and you're off the hook. There's not a jury in the world that would convict you.)

OUTDOORS: Dursban or Diazinon properly applied should do the job. Read the label.

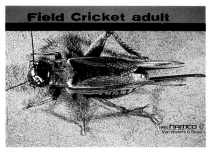

Field Cricket — adult
(VWR)

House Cricket — female and male
(VWR)

Jerusalem Cricket adult
(VWR)

60e.

Camel Cricket adult
(USDA)

Snowy Tree Cricket on tobacco
(USDA)

MOLE CRICKETS

Mole Crickets are nocturnal insects and are rarely observed during the day. The adults are powerful, though clumsy, fliers, but most of the time they remain in tunnels in the soil. Feeding on plant roots can occur, but that damage is considered secondary. Adults and nymphs damage young crops, grasses and tree seedlings primarily by tunneling in, under and around the root systems, and this thoroughly disrupts the roots' environment. The tunnels are made just under the soil surface, are about ½ inch in diameter and look very much like miniature ground mole (animal) galleries.

Mole crickets are about 1½ inches long, golden to chocolate brown in color, and covered with fine velvety hairs. They are near-cylindrical in shape with husky, mole-like digging forelegs. Although several species are found in Florida, the two most destructive are the Tawny mole cricket and the Southern. The Northern is the only native species.

Most mole cricket feeding occurs at night during warm weather, after rain showers or irrigation. Both adults and nymphs come to ground surface at night to search for food. Tunneling of more than 20 feet per night has been observed. Adults are strongly attracted to lights during their spring dispersal flights, and that's why you are likely to see them on your lighted front porch or in your lighted swimming pool.

Tawny Mole Cricket = Order: Orthoptera; Family: Gryllotalpidae
Southern Mole Cricket = Order: Orthoptera; Family: Gryllotalpidae
Northern Mole Cricket = Order: Orthoptera; Family: Gryllotalpidae

CONTROL CLUE

Mole crickets are very cannibalistic . . . and it couldn't happen to more deserving critters. Also, when they surface to feed, they become supper for fire ants, ground beetles, certain spiders, raccoons, skunks, red foxes and armadillos. So that's two points for our side, but your best bet for dependable control will be Oftanol granules or a Dursban or Baygon bait. Early June for Oftanol and late June for the baits is the optimum time for an application in north and central Florida; probably mid-May with Oftanol and early to mid-June for the baits in south Florida. Apply when overnight temperatures are 60 degrees Fahrenheit or above. If in doubt, call your county agent.

Southern Mole Cricket
(DOW)

Mole Cricket tunnel evidence in garden
(DOF)

Northern Mole Cricket
(VWR)

Left: Tawny; Right: Southern Mole
Crickets
(IFAS)

Curculios

COWPEA CURCULIO

The Cowpea Curculio might have a fancy name, but it's still just a cotton-pickin' weevil. Adults are about ¼ inch long, black, and possess that distinctive weevil snout . . . the big schnozzola. The larvae (grubs) are about ¼ inch long and are legless, white and C-shaped. Damage is done primarily by the grub with its well-developed chewing mouth parts.

These critters like beans — string beans, lima beans, wild beans, southern peas or cowpeas, but if times get tough, they'll even go after cotton seedlings. Here's how they do their thing. Mama curculio pierces a developing bean pod with her long snout and deposits eggs. Hatching occurs in about three days and grub-baby begins eating beans. The entire life cycle, from egg to adult, may be completed in 30 days. And once they start this jazz, kiss your beans goodbye.

Cowpea Curculio = Order: Coleoptera; Family: Curculionidae

CONTROL CLUE

Like for weevils — you gotta get mama before she becomes maternal and lays her eggs. Use Diazinon.

Cowpea Curculio attacking bean pod
(USDA)

PLUM CURCULIO

Plum Curculios are the reason you have wormy plums and peaches every year. Some damage is caused by adult weevils feeding, but the worst damage is caused by the larvae. Here's how curculios do their number on your fruit crop. When mama weevil gets the maternal urge, she will pierce a plum with that long snout and oviposit (lay eggs). At this point you can kiss that plum goodbye. When the eggs hatch, larvae (worms) develop and begin tunneling through the fruit, eating all the way. Left to their own devices, these grubs will eat more plums and peaches than you will.

Infested fruit usually falls to the ground, maturity notwithstanding. After the larvae complete their development, they pupate in the ground and become adults. These adults return top-side to repeat the carnage. Two or three generations of this pest can be produced in a single season, and they will begin their mischief early, even on a match-head size plum. One of the key things to look for is a characteristic C-shaped or crescent-like cut on the fruit. This scar is formed by the female during the egg-laying process. If you see a "C," you have a curculio problem. Stout-hearted gardeners fight back, and here's how you do it.

Plum Curculio = Order: Coleoptera; Family: Curculionidae

_____ **CONTROL CLUE** _____

Here's another case of *when* you spray being as important as *what* you spray. Use Malathion 50% E.C. (emulsified concentrate-liquid). Follow the directions on the label. When ¾ of the bloom petals have fallen, *spray*. This must be repeated every 10 to 14 days throughout the fruit growing season. The timing of the first spray will be critical. If you miss it, the remainder of your spray program will be mostly in vain. Once the fruit is infested, there is no way to kill the grubs inside the fruit. Oh, the price we mortals must pay for plums and peaches — without worms!

Plum Curculio and plum damage
(USDA)

Plum Curculio larva in plum
(USDA)

Cutworms

Here are some guys who can make vegetable gardeners cry. SUBTERRANEAN CUTWORMS feed almost entirely below the soil surface, eating roots and underground stems — Pale Western Cutworms are important members of this group. TUNNEL DWELLERS form and live in tunnels — stars of this group are Black Cutworms which will cut tender plants at the soil surface, pull them into their tunnels and eat 'em. Army Cutworms are night SURFACE-FEEDERS who spend the daylight hours hiding under soil, mulch or trash. Granulate Cutworms are other surface feeders that do serious mischief to garden vegetables in Florida. CLIMB-ING CUTWORMS, the athletes of the group, feed during the night and hide under boards and rocks during the day — Variegated Cutworms are good examples.

Oftentimes, the cutworm may be found during daylight hours (sleeping off his gluttony) curled up just below soil level in proximity to his latest conquest. If you find one, bust him with a rock, stick or anything handy. I have found he pops readily if squeezed between thumb and forefinger, but don't let your wife see you do this!

Cutworms = Order: Lepidoptera; Family: Noctuidae

CONTROL CLUE

A more civilized and perhaps more effective control would be applying a cutworm bait. Sprinkle the granules at the base of each seedling when you plant it. If the distasteful method (squeezing) or the civilized method (bait) are not appealing remedies, then you might want to try the old-time self-rising flour remedy. Sprinkle the flour, as you would the granules, around the seedling when you plant it. Mr. Cutworm does like that flour. He will come in the night and eat his fill. The next morning you will find a king-size cutworm, dead as a door nail. Caution! Use *self-rising* flour. He will thrive on regular flour and in a couple of days you will have raised the biggest, meanest cutworm "Go-rilla" in the neighborhood.

Black Cutworm and damage to corn
seedling
(USDA)

Climbing Cutworm and damage
(USDA)

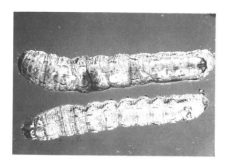

Variegated Cutworms in the morgue
(TAEX)

Earthworms

Earthworms in your garden? You should be so lucky! As they tunnel through the ground, earthworms ingest soil and in the process digest any organic matter in it. They usually deposit the soil that passes through their bodies as crumbly mounds (castings) on the soil surface. Although they are most numerous within the top 6 inches of soil, earthworms may tunnel as much as 6 feet down, bringing up deep layers of soil to the surface. These tunnels help to aerate and loosen the soil, which in turn improves soil drainage and tilth and facilitates plant root growth.

Earthworm activity in your ground is beneficial and should be encouraged. When I till my garden between seasons and uncover a bumper crop of earthworms, I know several things: my garden soil is healthy; humus content is good; moisture level is adequate; there is no active insecticide prevalent; and I'm on target for a successful next garden. I would rather suffer nematodes and grub worms than fumigate and destroy these benefactors. There is absolutely nothing wrong, however, in sacrificing a few down at the ol' fishin' hole.

I love earthworms, and so should you. Gardeners who consider them a nuisance don't understand Mother Nature . . . or fishin'.

Earthworms = Phylum: Annelida

_____ **BENEFICIAL** _____

Earthworm. Here's ol' good-buddy, but you can't tell whether he's coming or going.
(ORTHO)

Earwigs

Earwigs are hard, flattened, reddish-brown insects, up to 1 inch long, that are usually found in the garden. They feed on mosses, decaying organic matter, vegetation and other insects. They will also invade homes. They usually enter through cracks or openings in the foundation, or through doors, windows or other places. They are more apparent during hot, dry spells. In the home, they do not damage household furnishings, but their presence is annoying. From their abdominal glands, earwigs exude a liquid that has a disagreeable, tar-like odor. They are often found in dark, secluded places such as in pantries, closets, drawers and even in bedding.

Earwigs are only minor pests in the garden unless populations are high. They feed at night, and hide under stones, debris and bark in the daytime. Adult earwigs lay eggs in the soil in late winter to early spring. The young that hatch may feed on green shoots and eat holes in leaves, and as they mature, may feed on blossoms and mature fruit.

Earwigs become beneficial when they feed on other insect larvae and snails. Those frightening, forceps-like pincers that extend from the back end are used as offensive and defensive weapons with which they catch insects. Earwigs are also capable of giving you a good pinch if you mess with them, but the legend of their creeping into ears of sleeping persons is untrue.

Earwigs = Order: Dermaptera; Family: Forficulidae, Labiidae, Labiduridae

Earwigs — male and female
(VWR)

Because earwigs typically cause only minor damage in the vegetable garden, insecticide sprays are seldom needed. In the house, use an approved Diazinon spray. Bop him with a shoe if you don't mind that disagreeable odor.

Fabric Pests

CLOTHES MOTHS & CARPET BEETLES

Two types of clothes moths are common to Florida — CASEMAKING CLOTHES MOTHS and WEBBING CLOTHES MOTHS. Larvae of the casemaking clothes moth feed from a silken case which is dragged over the surface of their food; webbing clothes moth larvae feed within silken burrows which they spin over the fabric surface. Adult clothes moths are winged, buff-colored, and are about ½ inch long. They are weak fliers and avoid lighted areas. The larvae do the damage. They commonly feed on wool, feathers, fur, hair, upholstered furniture, leather, fish meals, milk powders, lint, dust, paper or even synthetic materials which may be soiled with oils; in other words, they will eat everything but the kitchen sink.

Adult CARPET BEETLES do not eat fabrics, but their presence in the springtime inside and outside the home should alert the homeowner. They may be seen crawling up walls of infested homes and also may be found congregating on windowsills. BLACK CARPET BEETLES are black with brown legs; COMMON CARPET BEETLES, FURNITURE CARPET BEETLES and VARIED CARPET BEETLES are mottled white, yellow, red and black. The body is usually covered with scales. Larvae of the black carpet beetle are carrot-shaped, covered with brown bristles, and have a tuft of long brown hair protruding from the end of the body. Larvae of the other species are short, stocky and are covered with brown and black bristles. Remember, the larvae do the damage by feeding on animal products such as wool, silk, hair, fur or feathers. They may feed for 9 months to 3 years, destroying carpets, felt padding, mohair cushions, furs and many other household items. These larvae can move from one type of food to another.

Casemaking Clothes Moth = Order: Lepidoptera; Family: Tineidae
Webbing Clothes Moth = Order: Lepidoptera; Family: Tineidae
Carpet Beetles = Order: Coleoptera; Family: Dermestidae

_____ CONTROL CLUE _____

The key to economic control of Clothes Moths and Carpet Beetles is good housekeeping. All furs, woolens, etc. should be sunned, brushed and dry-cleaned periodically. Sweep or vacuum regularly to remove lint, hair and dust from floors, shelves and drawers where adult clothes moths and carpet beetles may lay eggs and establish infestations. Give close attention to rugs, carpets, draperies, furniture cushions, closet corners, cracks, baseboards, moldings and other hard-to-reach places. Look for sources of infestation around the home such as old clothing, woolen scraps and yarn, furs, feather pillows and piano felts. It is vitally important that all cloth goods be dry-cleaned or washed, pressed with a hot iron, and brushed prior to storage in order to rid them of insects.

Do not apply sprays to furs; cold storage is the most practical method of damage prevention for furs. Cedar-lined closets and cedar chests have very limited value in fabric pest protection unless other measures are used. One pound of napthalene flakes or balls, or paradichlorobenzene (PDB) crystals per 100 cubic feet of closet space will provide adequate protection, but the area and contents will smell like "mothballs."

Rugs and carpet surfaces, especially around the edges and under heavy furniture, should not be ignored. Piano felt pads are best treated by a piano technician; upholstered furniture and pillows may require fumigation by a professional pest control operator because surface sprays will not control fabric pests inside the stuffing. Surface applications of Malathion, Diazinon or Methoxychlor should be made as coarse sprays along the edges of wall-to-wall carpeting in closets, corners, cracks, baseboards, moldings and other secluded areas. These treatments are particularly important for carpet beetles since they commonly live in lint and debris in these areas.

Are you sure you can do all this? Most good pest control firms provide dependable service for controlling Clothes Moths and Carpet Beetles; why don't you call one and get this job done right!

Casemaking Clothes Moths
(USDA)

Webbing Clothes moth
(USDA)

Webbing Clothes Moth damage to carpet
(USDA)

Black Carpet Beetle
(USDA)

Black Carpet Beetle damage to shirt
(USDA)

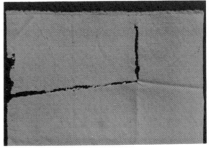

Furniture Carpet Beetle damage
(USDA)

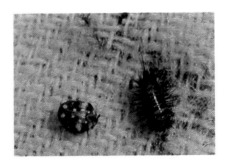

Furniture Carpet Beetle
(USDA)

Fleas

Fleas are found worldwide. There are an estimated 1600 species. The most common fleas encountered in Florida are probably Cat Fleas. Others, such as dog fleas, human fleas, oriental rat fleas and numerous other rodent fleas are encountered only occasionally by the homeowner. Female fleas in the family-way lay eggs shortly after a blood meal. Because fleas have the ability to survive for many months without food, they can remain in suitable areas for long periods of time waiting for dinner, which usually is your cat, your dog, or you. Occasionally eggs are laid while the female flea is on the host, but the eggs normally fall off. This is how homes become infested. Cat and dog fleas may also be found on humans, rabbits, squirrels, rats and birds. Human fleas also attack swine, goats, cats, dogs, rats, coyotes and skunks. Like, one big happy family!

Fleas = Order: Siphonaptera; Family: Pulicidae

Successful flea control must include treatment of the infested animals and also thorough treatment of the entire premises — indoors and outdoors. The effectiveness of pesticides is directly related to the thoroughness of applications. A number of insecticides are approved for specific uses. I'll give you a menu, but remember there are others. For the pet, ask your veterinarian; indoors, Dursban or Ficam; for the yard, Dursban granules. Maintain applications for 7 to 10 days to break the egg-hatch cycle. Or, call a qualified pest control service and be done with it.

Flea — magnified
(Ortho)

Cat Flea/Squirrel Flea comparison
(VWR)

Fleahoppers

When you hear conversations about Fleahoppers in Florida, it is likely to be among panhandle cotton farmers complaining about the damage these pests are responsible for in the cotton patch. The one that you are likely to be concerned with is the Garden Fleahopper. Male adults are winged and so are the females, but her wings are shorter. Both are nearly black in color. Fleahoppers have large hind legs enabling them to hop actively, which they do. The jumping habits and the hosts of the species have probably suggested the common name. Nymphs look like adults, except they are smaller, are greenish in color and are wingless. Fleahoppers attack southern peas, beans, cabbage, tomatoes, other garden crops and weeds. They are sap-suckers, making small discolored areas on the foliage. They can kill leaves, thereby injuring plants seriously. As many as five generations in one season have been recorded.

Fleahoppers = Order: Hemiptera; Family: Miridae

_____ CONTROL CLUE _____

Spray or dust plants with Malathion or Sevin. Fleahoppers aren't difficult to control. Certain weed hosts are #1 on their menu, so a good weed control program in and around your garden will be a big help.

Cotton Fleahopper
(TAEX)

_____ Flies _____

There are a great many species of Flies; some are dangerous to man as carriers of diseases and some are destructive to crops. But some are useful scavengers that clean up dead animals and plant wastes; others are insect destroyers — either predators like the Syrphid Flies or parasites like the Tachinid Flies that live in or on harmful insects. The larvae, called maggots, are footless, grub-like creatures that are usually soft, white or yellowish, with a head that is usually tucked away from view.

HOUSEFLIES are common throughout the world. In addition to being annoying, they can spread a number of serious human diseases and parasites such as diarrhea, dysentery, typhoid, cholera, intestinal worms and Salmonella bacteria. Flies feed on and lay their eggs in decaying organic matter; the eggs can hatch within 12 hours under ideal conditions. The creamy-white maggots burrow into and feed on the decaying material, pupate and then emerge as adult flies. The entire life cycle may be completed within 14 days. Several other fly species may infest the home, including FACE FLIES and LITTLE HOUSE FLIES.

FRUIT FLIES do not constitute a serious health menace, but can be annoying where fruit, vegetables or garbage is allowed to rot and ferment. The adult flies lay their eggs in the decaying fruits or vegetables. The eggs hatch in a few days, and the tiny maggots feed on yeasts growing in the decaying food. These tiny (up to 1/6 inch) VINEGAR FLIES are yellowish-brown, clear-winged and fly in a slow hovering manner.

Other fruit flies of extreme economic importance in agriculture are MEXICAN FRUIT FLIES, MEDITERRANEAN FRUIT FLIES and CARIBBEAN FRUIT FLIES. The Mexfly and the Medfly are subject to quarantine laws and continuous detection programs are monitored by state and federal agencies in order to keep those "biological-atomic-bombs" under control. The Caribbean fruit fly has now developed into the main fruit fly problem for citrus and several other crops in Florida. It now occurs in most of southern peninsular Florida, commonly north to Citrus and Volusia counties, with isolated records north to Jacksonville.

HORSEFLIES and DEERFLIES attack humans and domestic animals in rural and suburban areas. Female flies deposit their eggs in still pools of water, in moist soil or on vegetation. The larvae feed on decaying vegetation or on other insects and pupate in damp plant debris. The adult flies inflict painful bites that often continue to bleed after the fly has left. The human victim may suffer from fever and general illness. If ever you have seen an ol' cow standing out there, peacefully chewing her cud, then suddenly going berserk with tail flying and head flailing, chances are she has just had a hunk of her neck or rear bitten off by a horsefly or deerfly. The fly's mouthparts are like miniature scissors and a bite is . . . OUCH!

BLACK BLOW FLIES, BLUEBOTTLE FLiES, BRONZEBOTTLE FLIES and GREENBOTTLE FLIES are attractive devils, but there the compliments stop. They are thoroughly obnoxious otherwise. They lay their eggs on dead animals, garbage, sewage, or in open wounds of animals. Some related species parasitize and kill animals and even man. Eggs hatch very soon after being laid; larvae are mature in less than two weeks. The short life cycle means several generations a season. Their presence inside a residence sometimes indicates that a rodent or bird has died in a crawl space, wall void, chimney flue or attic. Blow flies are strongly attracted to odors of raw or cooked meat, poultry or fish. They are a nuisance indoors as they buzz around the room and bump into windows while trying to get outdoors.

FLESH FLIES are often a problem outdoors. They usually do not come indoors because the adult females prefer to deposit live larvae (no eggs) on meat scraps and dog excrement; therefore, they prefer areas around dog kennels and runs. Flesh flies are similar to houseflies, but they are slightly larger and have characteristic gray and black checkerboard markings on the abdomen or tail section.

HUMPBACKED FLIES are about ⅛ inch in length with a small head and large thorax causing the "humpbacked" appearance. They breed in decaying vegetation, animal debris and garbage, and also in ant and termite nests.

DUMP FLIES are about ¼ inch long and are black in color with a shiny thorax and abdomen. They are real "stinkers" that breed in garbage and fowl excrement, but they possess one redeeming attribute — the larvae are predaceous on larvae of other flies.

Housefly, Little Housefly, and Face Fly = Order: Diptera; Family: Muscidae
Fruit Flies = Order: Diptera; Family: Tephritidae
Horseflies & Deerflies = Order: Diptera; Family: Tabanidae
Blow Flies & Bottle Flies = Order: Diptera; Family: Calliphoridae
Flesh Flies = Order: Diptera; Family: Sarcophagidae
Humpbacked Fly = Order: Diptera; Family: Phoridae
Dump Fly = Order: Diptera; Family: Muscidae
Vinegar Flies = Order: Diptera; Family: Drosophilidae

_____ CONTROL CLUE _____

A combination of sanitation efforts, insecticide spraying and fly bait use should effectively control flies. The invention of cheap, mass-produced fly screening was one of man's greatest achievements toward assuring relief from flies. Indoor fly control should be 99% exclusion by employing screening, caulking, etc.

If large numbers of flies are present in your yard, then there is probably a fly breeding site on your property or nearby. Look for areas where flies may be breeding; clean up these areas and make whatever changes necessary to prevent this condition from recurring. If the source of the flies is not on your property, try to locate it and have the responsible persons help solve the problem. If cooperation cannot be obtained from the other person, contact your municipal or state health department to seek their assistance. Do not suffer with a serious, persistent fly problem that results from someone else's neglect.

The first step to control any fly problem successfully is to determine where the fly maggots are breeding. This process, termed source reduction, is always the most efficient method of control. Since flies feed and lay eggs in areas such as garbage cans or dog kennels which are sources of food odors, cleaning these areas eliminates the problem. Also control low fly numbers by using a residual insecticide spray to treat surfaces where flies usually rest. Dursban, Cygon or Baygon sprays are all good. Read the label. For fast knockdown, but with no residual control, aerosol fly sprays containing synergized pyrethrins, tetramethrin or resmethrin are worthwhile. These controls are appropriate for fast results and are safe applications around people, pets and food before or during picnics and outings. Vapona strips are worthwhile; many restaurants use them effectively. Why don't you get a couple?

House Fly
(VWR)

House Fly — the whole *#? family
(USDA)

Common House Fly/Lesser House Fly comparison
(VWR)

Oh, what big eyes you have, Face Fly!
(USDA)

Vinegar Fruit Fly larvae in fruit
(VWR)

Vinegar Fruit Fly adult
(VWR)

Mediterranean Fruit Flies on citrus
(VWR)

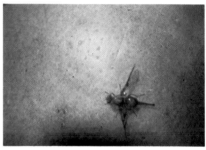

Mexican Fruit Fly doing her mama duty
on a grapefruit
(TAEX)

Black Deer Fly
(USDA)

Horse Fly adult
(VWR)

Blue Bottle Fly adults
(VWR)

Carribean Fruit Fly adult on guava
(DPI)

Carribean Fruit Fly maggot in guava
(DPI)

SYRPHID FLIES

I would like you to meet the Syrphid Fly. Now, here's a guy who ain't mad at nobody. He doesn't bite. He doesn't sting. He doesn't chew up leaves or eat anybody. He spends his time hovering over the flowers, sipping nectar. Adults are rather attractive black-and-yellow flies ranging up to ¾ inch in length. They rather resemble bees, but with a major difference. A syrphid fly has two wings, a bee has four. As the syrphid fly indulges in his nectar nipping, he inadvertently does a super job of pollinating. They are also attracted to tree sap and fermenting fruit.

Syrphid fly kids, you won't take to. They are slimy, green maggots. Although these things are as ugly as sin, they do a super job of eating aphids and they perform this chore in a grandiose manner. They have pointed jaws with which they grasp the aphid, raise it into the air and suck out all the body contents, leaving the empty skin. Syrphid fly larvae are usually found on leaf surfaces, often in the midst of a colony of aphids, chomping away. A single larva is credited with destroying an aphid a minute over long periods of time, and there may be many larvae in a garden. Tolerate them, slime and all. They are your friends.

Syrphid Flies = Order: Diptera; Family: Syrphidae

_____ **BENEFICIAL** _____

Syrphid Fly adults on flowers
(VWR)

Syrphid Fly larva feeding on an aphid
(VWR)

TACHINID FLIES

Tachinid Flies are beneficial insects; and because these flies are prolific, their value as parasites is increased. More than 1,400 species have been described. All species of the family Tachinidae are parasitic on other insects ... what a family of good-guys. Most Tachinid Flies resemble overgrown houseflies; they are grey or brown or black, but without bright colors. Eggs are usually glued to the skin of the host insect, but are sometimes laid on foliage where the insect will eat them along with the leaf. The larvae feed internally on their hosts which almost always die. One species parasitizes over 100 different caterpillar pests. Common prey are hornworms, cutworms and armyworms. Long Live the Tachinid Fly ... Long Live the Tachinid Fly ... God Save the Queen! Cheers ... Cheers ... Cheers.

Tachinid Flies = Order: Diptera; Family: Tachinidae

The West Coast Lady tachinid fly parasite

Van Waters & Rogers
1980 division of Univar

Tachinid Fly doing a number on a horn worm
(TAEX)

Tachinid Fly adult and pupa
(VWR)

_____ Galls, Insect-Induced _____

Galls are abnormal swellings of plant tissue caused by mechanical injury (autos, lawnmowers, kids, etc.) or by certain insects, bacteria, fungi, mites or nematodes. Among the insects causing galls are certain moth caterpillars, beetles, flies, psyllids (jumping plant lice), aphids and small wasps. Insect-induced galls are the most common galls in urban areas.

It is important to point out that gall tissue is a plant-product formed in response to a specific stimulus received from an insect, and once formed, these growths do not continue to utilize host plant nutrients; they are not parasitic. Galls are found most commonly on stems and leaves, but also occur on flowers, fruits and on the trunk. Evidence to date suggests that insect-induced galls in Florida do not result in significant plant injury. The exception is found in the occasional damage caused by phylloxera on pecans. Also, when plants afflicted with galls are refused by customers, the economic loss is very real to a nursery owner.

Gall-making Aphids = Order: Homoptera; Family: Aphididae, Coccidae, Psyllidae
Gall-making Flies = Order: Diptera; Family: Cecidomyiidae, Tephritidae, Agromyzidae
Gall-making Wasps = Order: Hymenoptera; Family: Cynipidae, Tenthredinidae
Gall-making Beetles = Order: Coleoptera; Family: Buprestidae, Curculionidae, Cerambycidae

Unless chemicals can be applied when gall-inducing insects fly or when adults are actively depositing eggs, they offer no effective means of control. Systemic insecticides have not proved effective. Once the gall begins its development, it is impossible to stop or reverse its growth with chemicals. In the case of ornamental oaks, the use of sprays should be avoided during late spring and mid-summer, since beneficial parasitic wasps are primarily active during these two periods. In the case of persistent branch and trunk galls, emergence of the adult gall-insect leaves a cavity that is subsequently occupied by beneficial critters. For example, when the adult emerges from the mealy oak gall, the vacated space is taken over by small spiders, lacewing larvae, ants or beneficial wasps. Thus, old galls provide protection for some good guys, which in turn eat the bad guys. In other words, find something else to worry about . . . unless you sell plants.

Wasp gall on oak leaf
(DOF)

Leaf gall on camelia
(USDA)

Galls on pecan caused by Phylloxera
(TAEX)

"Berry corsage" on oak caused by wasp
(DOF)

Leaf gall on azalea
(USDA)

Midge gall in slash pine needle bundles
(DOF)

Wasp gall on oak shoots
(DOF)

Mealy oak gall wasp ovipositing into oak bud
(VWR)

Cynipid wasp gall on water oak
(USDA)

Stem gall on cottonwood caused by petiole gall aphid
(DOF)

Midge leafspot galls on hickory
(DOF)

Cypress twig gall midge
(DOF)

Ocellate maple leaf gall
(USDA)

_____ Grasshoppers _____

Grasshoppers feed on a wide range of plants including vegetables, but of approximately 600 species occurring in the United States, few are of economic importance. Periodically, when conditions are right, grasshoppers increase to tremendous numbers and devour every green plant in their path. This occurrence is not the rule, but the exception. Early in the Fall of 1983, a horde of grasshoppers invaded downtown Houston, Texas. They swarmed over several of the big glass buildings, apparently seeking warmth and shelter. They made a big mess, but were hardly an agricultural threat.

Several grasshopper species will feed on rose bushes. They will devour not only leaves of the plant, but also buds, flowers and stems. Just a few grasshoppers can cause severe damage in a very short time, so act quickly if you see them around cherished plants.

Short-horned Grasshoppers = Order: Orthoptera; Family: Acrididae
Long-horned Grasshoppers = Order: Orthoptera; Family: Tettigoniidae

_____ **CONTROL CLUE** _____

Since grasshoppers migrate, control might be difficult. If you have a prize rose in bloom, cover it with netting 'til the grasshoppers are under control. If you are forced to use chemical control on your vegetables or roses, use either Malathion or Sevin.

Eastern Lubber Grasshopper
(DOW)

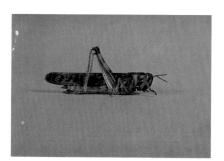

American Bird Grasshopper
(DOF)

124

Red-legged grasshopper and damage
(USDA)

Grasshopper feeding damage
on eucalyptus
(DOF)

Out-of-town grasshopper
photographed while attending a
convention in Miami
(TAEX)

_____Ground Pearls_____

Ground Pearls . . . might make a nice necklace for the wife, right? Wrong! These are mealybugs that live in the soil and suck juices from grass roots. The adult female is 1/16 inch long, pink in color, and has well developed forelegs and claws; adult males are tiny and gnat-like. Eggs are laid in the soil from March to June. The life cycle from egg to adult requires at least one or possibly two years.

Immatures are spherical, ranging in size from a grain of sand to about ⅛ inch diameter. They are yellowish-purple in color and actually look very much like pearls. Ground pearls ordinarily infest centipedegrass in north and northwest Florida, but will also infest all other Florida turf grasses. Severely infested turf turns yellow, then brown.

Ground Pearls = Order: Homoptera; Family: Margarodidae

Ground Pearls among plant roots
(DPI)

_____Harlequin Bug _____

The Harlequin Bug is a stink bug dressed up for Saturday night. Adults are red-and-black spotted, are about ⅜ inch long and are usually mentioned in the same breath with stink bugs because they do the same mischief. These plant bugs will congregate in large numbers on a host plant and literally suck it to death. This Pretty Boy Floyd of the insect world is guilty of everything except bank robbery. He will feed on the fruit of a wide range of plants including beets, squash, beans, peas, tomatoes and corn, causing shriveling and deformity. Don't let the fancy dress mislead you; this guy is a gangster.

Harlequin Bug = Order: Hemiptera; Family: Pentatomidae

_____ **CONTROL CLUE** _____

Remember, if you see harlequin bugs, damage won't be far behind, so hit 'em quick with Sevin spray or the Pyrethrins spray that you used on the other stink bugs.

Harlequin Bug on Chinese cabbage
(DOW)

_____ Head Lice _____

AND OTHER UNMENTIONABLES

Note: The following page is rated XXX and should not be read if nosy neighbors are over for coffee!

Oh, yes . . . it *can* happen to you. Head lice can happen to anyone. It is not a matter of poor health habits or of being dirty. Head lice are usually transmitted from one infested person to another by direct contact with the hair. Personal items like combs, brushes, towels and bedding are other frequent sources of contamination. Another source of infestation is clothing such as hats, ribbons, scarves, topcoats and sweaters.

Head lice do not observe any class distinction; they can infest anyone and are a problem which is rapidly gaining ground in all strata of our society. The schoolgirl who borrows a comb . . . the student trying on a hat in a fashion shop . . . children who share a bed for a nap . . . the traveller resting his head against the back of an airline seat. That's all it takes. Teachers in our elementary schools are advised to take note of children who immoderately scratch themselves. Head lice are epidemic in our schools. It is a national problem.

Human Sucking Lice = Order: Anoplura; Family: Pediculidae

CONTROL CLUE

Although lice are difficult to see, they are easy to recognize provided you're looking for them. Teachers, school nurses and hawk-eyed mothers will readily spot the small, silvery eggs attached to individual hairs. These eggs are called nits. In checking the scalp, pay particular attention to the back of the head and behind the ears. And, mother, if li'l darlin' comes home from school with a note stating he/she has flunked the cootie test, don't crater. It's not the end of the world. Your druggist or your family doctor can tell you what to do. And do it, or your whole family will go down scratching.

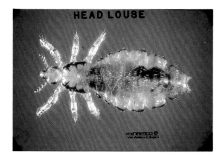

Head louse
(VWR)

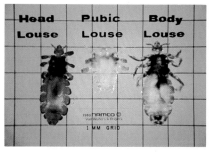

Head louse/Pubic louse/Body louse
comparison
(VWR)

Head louse nits on hair
(VWR)

Head louse nits on plastic comb
(VWR)

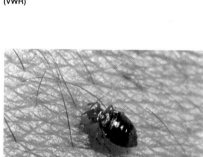

Bedbug
(USDA)

Bedbug nymph engorged after a blood meal
(VWR)

Body louse eggs in seams of clothing
— magnified
(VWR)

Body lice and eggs
(VWR)

Hickory Shuckworm

Hickory Shuckworms frequently cause severe injury to pecan and hickory crops in Florida. Little white worms with brown heads (larvae) infest the shucks covering the nuts. In later summer and fall, the shucks are tunneled out and as a result of this action, nuts are slower to mature; kernels do not develop properly; and shucks stick to the nuts and fail to open. Larvae overwinter in the fallen pecan or hickory shucks. They pupate in late winter and early spring, emerging as adults during spring and summer. Adult moths, which are dark brown to grayish-black and are about ⅜ inch long, deposit their eggs mainly on leaves and young nuts. The hatched larvae feed in developing nuts in early summer. Succeeding generations develop in pecan nutlets during June, July and early August and in the shucks during the remainder of the season. As many as 5 generations may be completed each year before fall larvae go into hibernation.

Hickory Shuckworm = Order: Lepidoptera; Family: Tortricidae

CONTROL CLUE

Good housekeeping will go a long way toward controlling this pest. Larvae cannot mature in decaying shucks, and adults cannot emerge from the soil. So, during July and August, rake fallen shucks and bury them. Be sure one of your Zolone sprayings occurs in proximity to August 15th. You can readily control this critter if you pay attention.

Hickory Shuckworm larva
(USDA)

Hornworms

The Tomato Hornworm and the Tobacco Hornworm, though genetically different, are so alike in appearance and dirtywork, I thought we might save time by considering them as one. First of all, these hornworms come from truly magnificent parents . . . large, fast-flying hawk moths (sometimes mistaken for hummingbirds) that are endowed with a 5-inch wingspan. And that's the only nice thing we can say about hornworms.

For the record, the tobacco hornworm larvae have seven diagonal light stripes; the tomato hornworm larvae have eight curved stripes. Basic colors are green and brown with a few shades in between. A red or black "horn" projects from the rear end. And can they eat! Hornworms feed on the fruit and foliage of tomatoes, peppers and eggplant. A few large larvae can strip a plant before you know it; if allowed to do so, they will grow to the size of a good cheroot cigar.

Incidentally, sometimes these worms may be found with puffed-rice-like white sacs on their bodies. These sacs are the cocoons of parasitic wasps (Braconid Wasp) that feed on and will eventually kill the hornworm. Better you should let the wasp take care of this hornworm and in the process make another batch of parasitic wasps to prey upon more hornworms.

Hornworms = Order: Lepidoptera; Family: Sphingidae

Tobacco Hornworm on tomato
(USDA)

Tomato Hornworm/Tobacco Hornworm/hornworm damage to leaf
(ORTHO)

Your best bet (other than the wasp): pick 'em off the plant, drop 'em on the ground and step on 'em.

_____ Lace Bugs _____

Lace Bugs are small, broad, flat little dudes that damage several broadleaf evergreens such as azaleas, laurels, pyracanthas and rhododendrons. Deciduous trees (those that shed their leaves) which may be attacked are sycamore, oak, hawthorn, hickory and ash. Lace bugs get their name from the appearance of the area behind the head and from the wing covers which are generally lacy-transparent. The lace bug nymph (baby) is flat, black, spiny and oval-shaped — looks nothing like mom and dad.

Adult lace bugs, which are about ⅛ to ¼ inch long, become active in the spring; after mating, females lay eggs on the ribs of leaves, usually the undersides. Nymphs emerge and feed on the leaf undersides along with the adults. Both have piercing-sucking mouthparts they use to suck out the contents of plant cells. There may be 3 or 4 generations a year.

Damage appears on the upper leaf surface as whitish, yellowish or bronze blotching. The presence of shiny black spots of excrement on the undersides of leaves is another good indication of lace bug infestation. A good case of lace bugs will also cause early leaf drop. If you have a sycamore tree and it begins dropping leaves in August, check for a sycamore lace bug infestation.

Lace Bugs = Order: Homoptera; Family: Tingidae

_____ CONTROL CLUE _____

If only a few lace bugs are present and little or no damage is observed, wash them off with a strong stream of water from the garden hose. If chemical control is necessary, use Orthene, Sevin or Malathion.

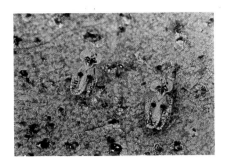

Lace Bug adults
(USDA)

Lace Bug nymphs on underside of
sycamore leaf
(DOF)

Lace Bug — magnified
(VWR)

Stipple-like Lace Bug damage to
sycamore
(DOF)

Lacewings

Here is Miss America of the bug world. In the adult form, she's one of
the most beautiful and is beneficial to boot. Adults are up to ¾ inch long,
have long antennae and transparent, lacy wings. Lacewings come in
several models, generally white, green or brown. Adults don't do a great
deal of anything except look pretty, make love, and eat a few aphids here
and there, but perhaps their most significant contribution lies in their ugly

progeny. The larvae are such voracious eaters that they are commonly called Aphid Lions. They suck the body fluids not only from aphids, but also from mealybugs, scales and other small soft-bodied insects that get in their way. Mama must lay her eggs singly on top of a delicate hair-like stalk so the bigger kids don't eat 'em. The larvae are rather small (up to ½ inch), are flat and rather resemble miniature alligators without tails.

If you're blessed with lacewings, you are most likely to find them on your screen door at night if the porch light is on. The adults are attracted to bright lights. They flop around as though they were just learning to fly. Don't spray them with bug stuff. Remember, they are good for you.

Lacewings = Order: Neuroptera; Family: Chrysopidae

_____ **BENEFICIAL** _____

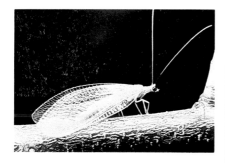

White Lacewing
(USDA)

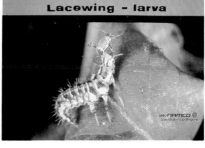

Lacewing larva (aphid lion) feeding on an aphid
(VWR)

Green Lacewing adult
(VWR)

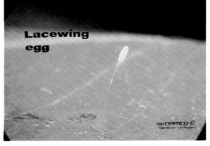

Green Lacewing egg on silk stalk
(VWR)

Brown Lacewing adult
(VWR)

_____Leaffooted Bug_____

No big deal about Leaffooted Bugs; they are stink bug kinfolks. What's unique about these cowboys is their odd appearance. These large (¾ inch) bugs are generally gray-brown colored with a distinctive white line across the back. Those strange flattened areas on their hind legs might remind you of someone wearing snow shoes. A leaffooted bug walks like a tomcat stepping through high, wet grass. They will put those yellow zits on your tomatoes and will pig-tail your okra, so don't ignore them.

Leaffooted Bug = Order: Hemiptera; Family: Coreidae

_____CONTROL CLUE _____

Any good vegetable garden spray containing Pyrethrins is what I use. It usually comes ready-mixed and can be applied even on the day of harvest. Read the label.

Leaffooted bug adult
(USDA)

Big-legged bug adult — another
member of the family
(DOF)

Leaffooted bug hind leg with those
characteristic cowboy "chaps"
(VWR)

___ Leafhoppers ___

Leafhoppers attack nearly all cultivated plants; most are wedge-shaped, elongated individuals, broad at the head and pointed at the rear. They are often brilliantly colored — green, yellow, red, or blue, and may have vivid color patterns — solid, striped, spotted or banded. Many, however, are a rather drab-colored brown, tan, greyish or black. Size is variable. The majority of adults range from 1/20 to ¼ inch, but a few will reach ½ inch in length, such as our ol' buddy the Sharpshooter. Probably more than 1000 Leafhopper species occur in North America.

When leafhoppers are abundant, host plants show a lack of vigor, growth is retarded and the leaves become somewhat mottled or whitened in appearance; ultimately, they turn yellow, red or brown. Leafhoppers are sap suckers and feed mainly on the undersides of leaves. This feeding often produces a burned or scorched effect and causes the tips to wither and die. Infested plants appear to have been seared by bright sunshine or drought.

Adults lay eggs in the plant stem, buds or leaves. Leafhoppers often become quite abundant on pasture and turf grasses, particularly Bermuda grass. In feeding, both the adults and nymphs inject toxins which poison plant tissue. Several species transmit plant viruses. The Yellow-Headed Leafhopper is a proven vector of the virus that causes Pierce's disease of grapes. Lethal yellow disease is passed to coconut palms and others by a Sharpshooter. Leafhoppers are attracted to lights and often swarm in great numbers around house and street lights, thereby becoming a nuisance.

If you walk through an infested area, large numbers of leafhoppers will hop or fly a short distance ahead of you, then drop down and hide. The Sharpshooter is perhaps most noticeable because he's larger than most of his leafhopper kin. He's the clown that plays hide-and-seek with you. He moves like a cursor on a computer screen, running backwards and sideways as rapidly as forward.

Leafhoppers = Order: Homoptera; Family: Cicadellidae

_____ **CONTROL CLUE** _____

Unless Leafhoppers are unusually abundant, insecticide control is not warranted.

I'm certain you've seen these critters. Don't count on "bugging" them to death; I've never heard of one going insane. Spray Malathion or Diazinon if you decide you need insecticide control and don't forget to spray the underside of the leaves.

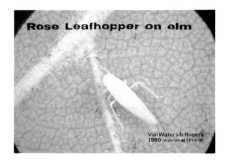

Rose Leafhopper adult on elm leaf
(VWR)

Blue and Green Sharpshooters
(VWR)

Pine Leafhopper adults. Look closely
— they made an "A" in camouflage
class.
(VWR)

Leafhoppers on a leaf
(TAEX)

Leafminers

Leafminers are the larvae of small flies, moths or beetles that tunnel between leaf surfaces. The two most common kinds, Serpentine and Blotch Leafmining Flies, are so named because of the shape of their mines in the leaves.

138

Leafmining Flies can be particularly damaging pests. They have the hots for beans, peas, celery, carrots, the crucifers, cucurbits, okra, potatoes and tomatoes as well as many other row crops. The adult is a small fly about the size of an eye gnat. She punctures the leaf and lays tiny eggs which hatch into small, yellow maggots which in turn feed on that chlorophyll tissue between the leaf surfaces. The result is that white, winding trail or pattern through the leaf.

You'll likely not want to eat the damaged portions of edible leaves, but it wouldn't hurt you if you did. (If you're not a picky eater, just consider that maggot an extra bit of protein.) Yields are usually not affected on fruit-producing vegetables unless populations reach epidemic proportions, which they periodically do. Generations in Florida are continuous for much of the year, but the epidemics occur more frequently in south Florida. Commercial vegetable producers can give you an earful on their woes with leafminers.

Leafmining Flies = Order: Diptera; Family: Agromyzidae

CONTROL CLUE

An ounce of prevention is about the only cure. Once this critter begins doodling on your tomato leaves, it's a little late to start a cure. About all you can do is get an early start with a Diazinon spray and keep your fingers crossed. Good luck!

Serpentine Leafminer
(DOF)

Blotch Leafminer
(DOF)

_____Leafrollers & Leaf Tyers_____

Leafrollers and Leaf Tyers are small caterpillars that feed inside leaf shelters which they make by rolling or tying together leaves. They are similar in their habits, but leafrollers roll leaves while leaf tyers tie leaves with silk thread. Simple. Larvae vary from pale yellow to dark green; all are about ¾ inch long when mature. Only part of their life may be spent feeding inside the rolled or tied leaves. At other times they may feed inside buds, flowers or fruit. Pyracanthas, carnations, chrysanthemums, geraniums, roses, zinnias, honeysuckles and verbenas are especially subject to attack, but some species will also attack fruit and hardwood trees.

The unique habit of rolling or tying those leaves together gives the insects some protection from unfavorable weather, predators and chemical sprays. Leafrollers and leaf tyers may be serious pests in the garden, feeding on many fruits, vegetables and ornamentals. Those species that feed on flowers and fruits are usually much more damaging than those that feed exclusively on leaves.

Leafrollers & Leaf Tyers = Order: Lepidoptera; Family: Tortricidae

_____ CONTROL CLUE _____

Check your plants periodically in the spring for the first sign of infestation; insecticides are most effective if they are applied before the larvae are protected inside the leaves. Use a spray containing Sevin or Malathion, but be sure your plant is listed on the product label.

Oak Leafroller larvae
(DOF)

Leafroller damage on pyracantha
(TAEX)

Lovebugs

Lovebugs are small black flies with red chests (thoraxes). Males are ¼ inch long; females are ⅓ inch in length. Why are they called Lovebugs? I just knew you would ask. These are the insects that prompt those embarrassing questions from the very young and envious comments from adults. Adult lovebugs are harmless and do not sting or bite. They feed on the nectar of various plants. Larvae feed on decaying plant material and perform a beneficial function by converting this material into organic components which can be used by growing plants. Two flights of lovebugs occur each year. The spring flight occurs during late April and May; a second flight occurs during late August and September. Flights extend over periods of 4 to 5 weeks. Mating takes place almost immediately after emergence of the females. Adult females live only 2 to 3 days, but I'm given to understand those are *happy* days.

Lovebugs are a formidable nuisance to motorists because they congregate in unbelievable numbers along highways and splatter windshields and grills of passing autos. They can clog radiator fins causing cars to overheat. They also get into refrigeration equipment on trucks, thereby causing overheating. The fatty tissue will cause pitting of a car's finish if it is not removed within a few days. They will soil your clothes and mess up fresh paint.

Lovebugs = Order: Diptera; Family: Bibionidae

CONTROL CLUE

Insecticidal control of the lovebug is impractical because infestations occur over such a vast area for so short a time. But if you think no control is tough luck ... how about those lovebugs! What a bitter end to a beautiful love affair ... being splattered all over the front end of an 18-wheeler!

Lovebug couple — illustration;
photographer sez he won't do no kind
of pornography.
(IFAS)

_____Mealybugs_____

Mealybugs are soft-bodied, sucking insects that are close relatives of scale insects. Plant parts heavily infested often appear to be covered with cotton; this appearance is caused by white or gray threads of wax with which mealybugs cover themselves. They are very active when young, crawling all over the plant until they find a suitable place to settle, but as the young mealybugs mature, they become sluggish. Mature females move around very little. Adult males do not feed, but die after mating.

Mealybugs may damage any part of a plant by sucking out the sap, which in turn causes leaf distortion, yellowing, stunting or galls, and ultimately can cause a plant to die. They also coat the plant with large quantities of undigested sap, called honeydew. A yeasty fungus then grows on the nutrient-rich honeydew which results in a black sooty mold mat that covers leaf surfaces and blocks photosynthesis. Most species of mealybugs are garden pests only in warmer climates, but will infest house plants and greenhouse plants in any climate.

Mealybugs = Order: Homoptera; Family: Pseudococcidae

Mealybugs aren't difficult to control if you can penetrate their waxy-threaded cover, so dump some spreader-sticker into your Malathion or Diazinon spray; that should get 'em. Spray again in 10 to 14 days, just to be sure. If you find only a few mealybugs on a house plant, put a drop of rubbing alcohol on each bug with a cotton swab. This will work nicely if you have the patience to do it.

Coconut Mealybug
(DPI)

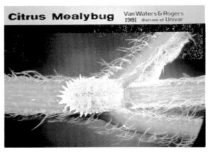

Citrus Mealybug adult magnified
(VWR)

Longtailed Mealybug infestation on orange leaf
(VWR)

Longtailed Mealybug nymphs, adults, wax and black sooty mold close up
(VWR)

Golden Mealybug infestation on Norfolk Island Pine close up with eggs, nymphs and adults
(VWR)

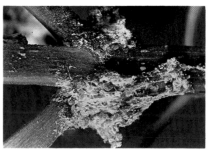

Mealybug infestation on coleus
(USDA)

Azalea Mealybug
(USDA)

Mealybug destroyer — adult Ladybird Beetle
(VWR)

_____Minute Pirate Bug_____

Not much need be said about the Minute Pirate Bug. He's a beneficial insect and I don't know anything bad about him. This little guy simply spends his time sucking on insect eggs, aphids and other small soft-bodied pests. He's not very ambitious, is he?

Minute Pirate Bug = Order: Hemiptera; Family: Anthocoridae

Minute Pirate Bug adult
(VWR)

_____Mites_____

GARDEN & LANDSCAPE MITES

Mites are not insects, but are more closely related to spiders and ticks. Most mite species feed not on plants, but on organic matter in the soil, on other mites, or on insects. However, several mite species do attack house and garden plants. The most common of these are Spider Mites and Eriophyid Mites. Both are so tiny, you're not likely to ever see them without magnification; what you will likely see is evidence of their presence. They feed by inserting a tiny sucking beak into plant tissue to withdraw plant juices and chlorophyll. This results in a small colorless or whitish spot, and in time the infected leaves become stippled-like in appearance, then become dry and drop off.

SPIDER MITES are oval-shaped and when magnified appear to be covered with long hairs. Some members of this family produce large quantities of fine webbing. A point of importance in understanding this pest's habits is knowing of its dispersal mechanism. When mites increase in number to the point there is competition for food, they will mass toward the top of the plant. Individual mites then drop on silken threads, forming a "rope" of living mites. Spread to new plants or locations can then occur by air movement, by human and animal contact or by cultural practices.

ERIOPHYID MITES, commonly called gall mites, rust mites or bud mites, are even smaller than spider mites. These cigar-shaped mites are difficult to see, even with the aid of a magnifying glass. Another difference . . . these little "dudes" produce no webbing.

Showing you pictures of a bunch of different mites would serve little purpose as a prelude to control, but here's a test for spider mites in the event you don't have a microscope. Hold a clean, white sheet of paper under the sick plant leaf. Briskly thump the leaf several times. You should see several minute specks on the paper. With a pen, draw a tight circle around each speck. Now, wait. If the specks move out of the circles, then they are alive. Welcome to Spider Mites. Don't sneeze while performing this test. The specks will disappear and you'll have to do everything all over again.

Spider Mites = Class: Arachnida; Order: Acari; Family: Tetranychidae
Eriophyid = Class: Arachnida; Order: Acari; Family: Eriophyidae

_____ CONTROL CLUE _____

Choice #1 — Kelthane; Choice #2 — Diazinon with wetting agent; Choice #3 — Malathion. Read the label. Under Florida conditions, mites complete their life cycle in 7 to 10 days at 80 degrees Fahrenheit, so spray again in 5 or 6 days.

Spider Mites with web and plant injury
(USDA)

Spider Mite damaged leaves
(TAEX)

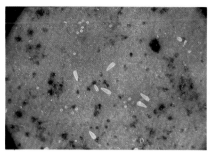

Citrus Rust Mite
(TAEX)

Spider Mite damage to holly
(USDA)

So you insist upon seeing a Spider
Mite. Okay, okay, already! Here.
(TAEX)

HOUSEHOLD & HUMAN MITES

WARNING: The Surgeon General has determined that reading this chapter will make you scratch!

Mites are occasionally found in homes and attack humans in the absence of their normal hosts — birds, rodents or insects.

BIRD MITES: The Northern Fowl Mite, Tropical Fowl Mite and Chicken Mite are the major bird mites in Florida. They are usually encountered in homes when they migrate from bird nests in eaves, rafters or gutters. When the young leave the nest, the mites will migrate to other areas in search of a blood meal . . . and that usually is *you*.

INSECT MITES: Straw Itch Mites are the most prevalent insect parasites that also attack humans. Their bites are located almost entirely on the clothed portions of the body; dermatitis results within 24 hours. Humans become infested when they come in contact with straw, grain or wood.

RODENT MITES: The Tropical Rat Mite and the House Mouse Mite are the two most abundant rodent mites found in buildings. They can cause severe irritation and dermatitis in humans. Areas bitten by these mites may remain swollen for several days and leave red spots.

SCABIES MITES: The Scabies Mite or Human Itch Mite burrows into the skin of humans causing human mange or scabies. Different varieties of these mites are specific for certain mammals including man, domestic animals and wild animals. Scabies mites are transmitted from one person to another by direct contact or also by two persons using one bed. Persons infested with scabies suffer intense itching; a rash may develop later around armpits, the wrists, the waist and back of the calves. Even though only a few mites may be present, the rash may spread over much of the body.

HOUSE DUST MITES: House Dust Mites (floor mites) are a little known and rarely mentioned group of mites of medical significance to humans. Their importance is that pieces of the mites in house dust may produce allergic reactions when inhaled. House dust mites feed on the shed skin of humans and are commonly found in houses and schools throughout the United States. You say you're allergic to housework? Could be!

Are you itching yet?

Mites (Human & Household) = Class: Arachnida; Order: Acari
Bird, Insect & Rodent Mites = Family: Psoroptidae
Scabies Mites = Family: Sarcoptidae
House Dust Mites = Family: Acaridae

_____ CONTROL CLUE _____

HOUSEHOLD MITES: Control of household mites is best accomplished by eliminating the nests and roosting areas of birds, by controlling rodents and by controlling insect hosts. Likely infested areas should be treated with an approved Dursban or Diazinon spray. Bites should be treated with antiseptic and a local anesthetic may be applied to ease the irritation. Persons with resulting severe dermatitis should see their physician for treatment.

SCABIES MITES: If scabies mites are suspected, see your family doctor for diagnosis and treatment. If you suspect that you might have picked up dog scabies from your pet (and that's possible), in addition, see your veterinarian.

HOUSE DUST MITES: No chemical control of house dust mites is presently known. Frequent bedding changes and use of non-fibrous bedding material will reduce mite populations; frequent vacuum cleaning and lessening of excess humidity will also help.

Rat Mite off a Norway rat
(VWR)

Scabies Mite or Itch Mite
(VWR)

House Dust Mite
(VWR)

Mosquitoes

Throughout history mosquitoes have occupied a position of importance as a pest of mankind. Worldwide, there are 1500+ mosquito species; 67 of these are pests, to varying degrees, in Florida. Mosquitoes are

149

responsible for the transmission of disease to millions of people in the world each year. These diseases include encephalitis, dengue, yellow fever, malaria and filariasis. Two domestic, disease-carrying species, *Aedes aegypti* and *Culex pipiens* and those domestic or semi-domestic species of *Anopheles* which carry malaria are understandably the most widely publicized, but in Florida, the big problems are mostly with such wild mosquitoes as *Aedes taeniorhynchus, Aedes sollicitans, Psorophora columbiae, Culex nigripalpus* and *Coquilletidia perturbans.*

When we define mosquitoes as wild or domestic, we are grouping them in broad categories which can be further narrowed. For instance, we may group the wild ones ecologically into woodland and field species, and these yet further into swamp mosquitoes, savannah mosquitoes, jungle mosquitoes, tundra mosquitoes and so on. If you're concerned with their biting habits, you'll divide them into non-biters, man-feeders, mammal feeders, bird feeders and so on down the line.

Mosquitoes are involved in a certain amount of Florida history. When the issue of statehood for Florida was debated in Congress in 1845, John Randolph of Virginia described the area as a land of quagmires, swamps, alligators and mosquitoes that would never amount to anything and could be forgotten. In 1850 the state had its first dengue fever epidemic and malaria was observed for the first time. These crises prompted Dr. John Gorrie of Apalachicola to invent the first ice machine in order to cool the fever of his patients. The city of Jacksonville is noted for the great pestilence of 1857 which was brought on by a malignant strain of yellow fever. Steamers would not stop and Jacksonville was isolated from the world. Six-hundred persons were infected with yellow fever; 127 died. I would imagine John Randolph's prophecy was recalled in 1888 when Jacksonville suffered one of its greatest epidemics. The city at that time had a population of around 20,000, and by Thanksgiving 5,000 people had succumbed to yellow fever; 400 had died.

It might be interesting to note several typical as well as atypical mosquitoes that occur in Florida:

Psorophora ciliata and *Psorophora howardii* are very large mosquitoes also known as gallinippers. They are formidable because of their size and vicious bites. *Psorophora columbiae* is commonly called the glades mosquito or pasture mosquito. *Columbiae* females are smaller than the *ciliata*, but they can make life unbearable for humans and have been known to kill cattle when present in large numbers. Gallinipper larvae are cannibalistic; they prey on *columbiae* larvae and the smaller larvae of their own species. These three floodwater species can occur in hordes of unbelievable numbers. They have been known to mass on a new-born calf and smother it. (Lawrence Rhodes, Director, Manatee County Mosquito Control District, told me these species get so numerous that surveyors in the everglades have to throw rocks in the direction of their rod men and quickly holler thru the hole to communicate. I'm sure he was teasing, but I got the point.)

Aedes aegypti, often called Tiger Mosquitoes, are perhaps the most domesticated and breed in areas where humans live. They lay their eggs singly on the sides of containers just above the water line to await the next rainfall to provide water to cover them. The female is a quiet attacker and usually likes to bite around a person's ankles or back of neck. She will bite during the day and in lighted rooms at night. This mosquito transmits yellow and dengue fever to humans and heartworms to dogs.

Culex nigripalpus lays egg rafts in artificial containers, highly polluted waters, roadside ditches, rain-diluted salt marshes and elsewhere. It's a night feeder and will take blood from birds, mammals and reptiles. This mosquito has been identified as a vector of St. Louis encephalitis in the south half of Florida.

Anopheles quadrimaculatus is the malaria mosquito. It breeds primarily in freshwater ponds and swamps. Although malaria has disappeared as a significant problem in the United States, it is still one of the most important communicable diseases on a worldwide basis. Malaria in man is an acute or chronic disease caused by any one of four species of protozoan parasites that can be transmitted from person to person by the bite of *Anopheles* mosquitoes.

Deinocerites cancer, the crab hole mosquito, lays eggs above the water line in land crab holes along the beach. Tidal percolation later floods the eggs and they hatch. This mosquito has become a significant pest along some beach areas.

Wyeomyia vanduzeei, the bromeliad mosquito, breeds in the water of both ground and tree borne bromeliads. It is a daytime biter and is a significant nuisance in south Florida where large oak trees may host thousands of bromeliad plants.

Coquilletidia perturbans and *Mansonia* species are unique in that larvae do not surface for air, but rather attach to aquatic plants via a modified air tube and take oxygen from the plants. The only larval control is herbiciding.

Uranitania saphirina is a rather exotic iridescent mosquito that takes blood mainly from reptiles.

Toxorhynchites rutilus is an extremely large, iridescent mosquito that feeds on nectar only; neither sex bites. The larvae are predacious on other mosquito larvae. Tests are being conducted to determine if they might be an effective control of tree hole and container mosquito species. AT LAST . . . a mosquito "good guy."

Mosquitoes = Order: Diptera; Family: Culicidae

The most effective mosquito control program will be performed by an entity like your local Mosquito Control District; however, individuals can contribute a great deal by practicing the Ten Commandments of Mosquito Control:

1. Empty, remove, cover or turn upside down any receptacle that can hold water — particularly old bottles and tin cans.
2. Change water and scrub vases that hold flowers or cuttings twice each week. (Or grow your cuttings in damp sand.)
3. Discard old tires or store them indoors.
4. Screen rain barrels and openings to water tanks or cisterns. Seal cisterns not in actual use.
5. Repair leaky plumbing and outside faucets; they may cause standing water.
6. Clean clogged roof gutters and drain flat roofs.
7. Fill holes in trees with sand or mortar, or drain or spray them.
8. Stock ornamental ponds with mosquito-eating fish.
9. Connect open waste-water drains to a sewerage system or construct a separate sump or leach.
10. Fix or replace screens on doors and windows.

Mosquitoes are attracted to people for a variety of reasons — by the carbon dioxide they emit or by the color of clothing or by perfumes and aftershaves, but the most pervasive reason is for that blood meal necessary for procreation. Dark or blue colors attract mosquitoes, but they seem to stay away from light colors (except blue). Mosquitoes, however, tend to swarm around someone dressed in white at twilight. Yellow light bulbs will have less tendency to attract some species that are attracted to light. Repellents are perhaps the best deterrent for an individual; liquid repellents can protect against mosquito bites for 2 hours or more depending on the person, species of mosquito that attacks and the number of mosquitoes. Other devices available for mosquito control — electric thermal foggers; engine driven thermal foggers; electric misters; hand-held portable insecticide foggers; and backpack insecticide foggers. Remember, in the effort to control mosquitoes ... we hang together, or we slap alone!

Mosquito (female adult) about to feed on human
(MCMCD)

Mosquito engorged after feeding on human. BURP!
(MCMCD)

Culex female laying eggs
(MCMCD)

Culex egg rafters
(MCMCD)

Mosquito larvae
(MCMCD)

Adult mosquito emerging from pupal case
(MCMCD)

Newly emerged adult female mosquito
(MCMCD)

Designer wear in Gulf Coast Salt Marsh
— boots and jeans a la mosquitoes!
(MCMCD)

Moths

CLEARWING MOTHS

Clearwing Moths are rather wasp-like in appearance. Common colors are yellow, black or dark blue. Adults are moderate in size, with the front wings being exceptionally narrow. And another striking characteristic is that the wings generally are relatively free of scales. This is uncommon to most moths. They are active during the day.

Mature larvae are white to purplish gray borers that make round holes in the bark of favorite trees which might be oaks, willows, cottonwood, elm or others. You might also notice sapstaining around the entry hole along with an accumulation of fine to gooey frass caught on branches or the trunk or at the base of the tree. Infestation may also cause foliage discoloration and/or wilt with resulting branch dieback. Larvae make shallow galleries underneath the bark.

It might make you feel better to know that it would take a "goodly" number of these borers to seriously threaten a tree. Keep your chin up!

Clearwing Moths = Order: Lepidoptera; Family: Aegeriidae

Treat for borers, because that's what these damage-dealing worms are, but since the galleries are typically shallow and clear, you might even be able to skewer this critter by forcefully poking a piece of straight coat hanger wire into the bark hole. And even if you don't stab this guy, you will at least give him a terribly sore rear end. If you happen not to be as sadistic as I am, you might also consider moderate doses of corrective pruning.

Exposed gallery of lesser dogwood borer damage to pecan
(DOF)

Sap staining and borer damage to tree from an invading Clearwing Moth
(DOF)

Borer damage close up
(DOF)

CODLING MOTH

Codling Moths are one of the two varieties of fruit moths that commonly attack deciduous fruits. Their behavior is very similar to that of the Oriental Fruit Moth, but the fruit of apples and pears are attacked more directly causing the fruit to drop early in the season. Pinkish-white, brown-headed larvae (worms), about ¾ inch long when fully grown, will tunnel into the core of the fruit and feed for several weeks. The entry holes are rather obvious because of a mess of brown, sawdust-like material surrounding them. Once invaded the fruit is ruined.

The codling moth is the most serious insect pest of apples, causing most of the wormy fruit which is so familiar to everyone. This insect is found wherever apples are grown. English walnuts and other fruits are also subject to attack. The warmer climate of Florida could prompt 3 or 4 generations annually.

Codling Moth = Order: Lepidoptera; Family: Tortricidae

CONTROL CLUE

Once the fruit has been penetrated by the larvae, it's too late to worry about control measures. This critter is tough, insecticides thoroughly applied at the correct time are the only way to go for effective control. I think you will need some help with this one. Call your county extension agent.

Codling Moth adult
(USDA)

DIAMONDBACK MOTH

Here's a little critter that causes some pretty big trouble. The adults are grayish moths about ⅓ inch long. Mom and pop aren't so bad, but again it's the blankety-blank kids (larvae) that do the damage. These little devils, which rarely exceed ⅓ inch in length, are pale yellowish-green caterpillars with fine, erect, scattered black hairs over their bodies. And they wiggle like the dickens if disturbed. They feed on the underside of leaves, leaving shot-hole type damage, and you go nuts trying to decide what's eatin' your plant. Usually the outer leaves are attacked. What plants are preferred? Cabbage, cauliflower, broccoli, and stuff like that.

Diamondback Moth = Order: Lepidoptera; Family: Plutellidae

CONTROL CLUE

Use Sevin, Dipel or Thuricide as you would for any other garden caterpillar, but be sure to spray or dust the underside of the leaves. You should always do that anyway.

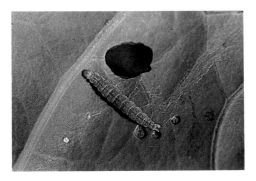

Diamondback Moth larva with damage
to leaf
(USDA)

GYPSY MOTH

The Gypsy Moth was brought from Europe to the United States in the spring of 1869 by Leopold Trovelot, a naturalist, for the noblest purpose of producing a commercial source of silk. It didn't work out that way. (Reminds me of the fellow who imported the English Sparrow.) Gypsy moths escaped from his house in Medford, Massachusetts, and in 1889 the first formidable outbreak of this awesome pest first commanded public attention. Current outbreaks are still limited principally to northeastern and middle Atlantic states, but the southward expansion seems to be as determined and devastating as Sherman's march through Georgia during the Civil War.

In Florida, potential damage is high to forests composed primarily of oaks and other favored tree species, particularly in portions of the panhandle and the central sand ridge of the peninsula. The gypsy moth is a proven defoliator of over 500 species of trees, shrubs and vines.

In the northeastern United States, moths emerge from late July to early August. Females emit a scent (pheromone) to attract males for mating and then lay eggs which are placed in dark, sheltered areas, e.g., bark crevices, under loose bark, and the undersides of limbs, rocks, stumps, vehicles and outdoor household equipment. Each egg mass may contain from 100 to 1,000 eggs. WOW! There is only one generation per year, but that's quite enough.

Gypsy Moth = Order: Lepidoptera; Family: Lymantriidae

CONTROL CLUE

A federal domestic quarantine and special control efforts are in effect for the gypsy moth in Florida. If perchance you wake up to worm-world one morning and the inhabitants are large (up to 2½ inches long), hairy, blackish caterpillars with two rows of longitudinal red and blue spots on their backs, friend, you got a problem. And it's bigger than both of us. Call your local county extension agent or your state forester for HELP! HELP! HELP!

Gypsy Moth caterpillar
(SFI)

Gypsy Moth adults —
female and male
(DOF)

ORIENTAL FRUIT MOTH

Oriental Fruit Moths perform the same sort of mischief as codling moths, only more so; the larvae feed on both twigs and fruit. Peaches, plums, apples and pears are the fruits chiefly attacked.

The larvae overwinter in cocoons under the tree bark. They feed on buds and succulent new growth in the terminals, creating a condition called "flagging" — the tips of damaged shoots break off. Later in the summer, successive generations attack the fruit directly when no juicy, young foliage is available. Mature larvae leave gum-filled holes in the fruit when they exit to pupate, and several generations can occur in a single year.

Oriental Fruit Moth = Order: Lepidoptera; Family: Tortricidae

_____ **CONTROL CLUE** _____

Worms in twigs and fruit cannot be killed with insecticides. It is the early-bird who gets this worm. Recent advances with pheromone traps make it possible to predict the egg-laying time of the mama oriental fruit moth. Check with your local county extension agent. If you can't do that, spray early with an approved formulation of Diazinon or Malathion when ¾ of the bloom petals have fallen. Repeat every 10 to 14 days.

Oriental Fruit Moth damage to peach twig
(USDA)

PINE TIP MOTHS

Pine Tip Moths are insects that feed inside the growing shoots, buds and conelets of young pines. The species most common in Florida are the NANTUCKET PINE TIP MOTH, THE SUBTROPICAL PINE TIP MOTH and the PITCH PINE TIP MOTH. A pine tip moth overwinters as a pupa inside a damaged shoot, cone or bud of a pine tree.

Adult moths emerge in the early spring, sometimes as early as February in Florida, when warm days become common. After emergence, the moths mate and the females lay their eggs on new pine shoots and conelets or on last year's shoots. In cool weather (late winter to early spring), eggs may take 30 days to hatch, but require only 5 to 10 days to hatch in hot weather (late summer). After hatching, young larvae, being young and daring, may feed on the outside of new growth for a short period of time. Later, being now older and wiser, the larvae bore into shoot tips, conelets and buds, feed within these tissues for 3 to 4 weeks and then moult into pupae. Adult moths then emerge, and the cycle begins again. In Florida, there may be 4 or 5 generations per year.

Adult pine tip moths are small and are rather attractive. They have gray heads and bodies. The front pair of wings have irregular patches of red, copper and gray bands. The young mischief makers (larvae) are cream-colored with black heads when young, but develop a light brown to dark orange body when older; they will be about ⅜ inch long. Pupae will be light to dark brown and about ¼ inch long. If you have young pine trees, be on guard for pine tip moths; they really get the hots for loblolly pine, shortleaf pine and Virginia pine. Slash pine is attacked infrequently; longleaf pine is rarely, if ever, infested. Most damage occurs during the first 5 years of growth, but don't bet money on that.

Watch for these symptoms of infestation: foliage discoloration — needles turn from green to reddish brown and will fall off; dead or dying branch tips, often curved or tipped; resin beads or flakes and fine silk webbing on branch tips (sometimes granular fecal material is present); damaged parts hollowed out — larva or pupa may be present.

Pine Tip Moths = Order: Lepidoptera; Family: Tortricidae

_____ **CONTROL CLUE** _____

Zolone or Orthene is a good foliar treatment; Furadan is a good soil treatment with systemic properties. Don't forget . . . read labels. If you raise Christmas trees for a living, contact your local forester about pheromone traps. I understand they work real good.

Nantucket Pine Tip Moth adult on loblolly pine needle
(SFI)

Pine Tip Moth damage to pine shoot
(DOF)

Subtropical Pine Tip Moth adult
(SFI)

Pine shoot dissected to show pupa of
the Nantucket Pine Tip Moth
(SFI)

Loblolly pine severely
damaged by Nantucket Pine
Tip Moth
(SFI)

Pine Tip Larva on pine shoot; note two
ectoparasites
(SFI)

WHITEMARKED TUSSOCK MOTH

The Whitemarked Tussock Moth is a common spring pest, especially around oak-populated areas. Central and northeastern Florida are notoriously susceptible to these critters. Common host trees include apple, elms, laurel oak, live oak, mimosa, pyracantha, redbud, water oak and other hardwoods. Larvae can cause severe defoliation of a host tree, and if this occurs for two or more consecutive years, death could result if that tree is otherwise stressed. The mischief-maker (larva) is a real "fancy-pants" with its coral red head, yellow to cream colored, black striped body and those exotic tufts of caterpillar hair.

Tussock moth larvae have a couple of other bad habits in addition to eating your tree. Young larvae disperse by falling or spinning out of trees on silk strands; this habit can be most unnerving if you happen to be standing under a tree watching this activity with open mouth. And another thing — those black dots on your glasses . . . in polite circles it's called falling frass, but in reality it's caterpillar poop. These critters are not toilet trained at all.

In high-expansion times, these guys will come in your house and may greet you in the bathroom or in bed between the sheets. Also, rash-like symptoms have been reported by some individuals exposed to repeated contact with those caterpillar hairs.

Enough is enough!

Whitemarked Tussock Moth = Order: Lepidoptera; Family: Lymantriidae

_____ **CONTROL CLUE** _____

Outside: spray these dudes with Orthene. Read your label! In bed or in the bathroom: you're on your own. Good luck!

Whitemarked Tussock Moth caterpillar
(DOF)

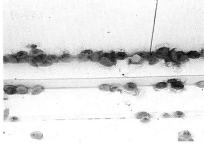

Whitemarked Tussock Moth cocoons and egg masses on house siding
(DOF)

_____Nematodes_____

There are some 80,000 varieties of Nematodes. Some 1500 of these are classed detrimental. Some nematodes are actually beneficial. A number of these "bad guys" attack many of your ornamental and garden plants as well as lawn grasses. Here's your nematode line-up by type: SPIRAL, AWL, ROOT KNOT, LESION, STUNT, FOLIAGE, STING, BURROWING and RENIFORM.

One of these, the Root Knot, you might identify by its symptoms, but most likely you will never actually see this nematode or any of the others unless you have a good microscope and know how to use it. Root Knot nematode presence is most noticeable to the observer because of that string of knots on the plant roots, but other nematode types cause similar plant damage with less obvious symptoms, such as stubby root condition, dead areas within roots, excessively branched roots or death of an entire root branch. And that's what you're up against when you have nematodes.

If your plants are legumes such as beans or peas, don't confuse those nitrogen-fixing nodules with root knot nematodes. Nitrogen nodules, attached to the side of the root, can be removed with the thumbnail without destroying the root, while root knot galls are formed within the root and cannot be removed without root destruction. Nematodes are capable of moving only about one foot per year by their own motion; therefore, they must rely on movement of soil, water or plant material for major distribution. Soil clinging to tillage equipment is one of the most common means of spreading nematodes.

Nematodes = Phylum: Nemathelminthes; Class: Nemotada; many orders and families.

CONTROL CLUE

Don't expect to eradicate nematodes by cultural or chemical means alone; however, a combination of suggested procedures can reduce nematode numbers so that a successful crop may be grown.

1. Plant nematode resistant varieties if available. 2. Tilling the soil regularly during summer, exposing it to heat and drying, to reduce populations considerably. 3. Rotate susceptible and less susceptible crops. 4. Don't rely on marigolds or put sugar in the soil. You can do more harm than good. 5. If nematodes get too bad, stop gardening in that spot for a year or two.

I don't feel that available chemical controls are worth the expense and the dangers incurred. Why don't you ask me something easy?

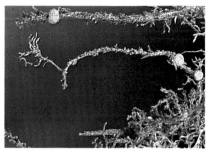

Sting or Lance nematode damage
(USDA)

Left: Root Knot nematode damage; galled roots; Right: Lesion nematode damage; sparse roots
(USDA)

Nematode damage to corn roots
(USDA)

Palm Leaf Skeletonizer

The Palm Leaf Skeletonizer is the larva of a small moth that does its thing only on palms. Both upper and lower surfaces of the leaf are subject to attack; large brown splotches develop in the leaf as the gregarious larvae feed. *The larva is the only destructive stage of this critter.* Infested palms are seldom killed, but dead leaf areas and webbed galleries incorporated with frass (mess) destroy the beauty of palms and diminish their utility as ornamental plants.

Eggs are deposited on the underside of the interleafing husk which surrounds the leaflets of the palm during early stages of growth. Incubation lasts 9 to 25 days. Young hatchlings feed on the leaf tissue directly under the old egg case. The larvae spin a web of silk under which the whole colony is protected as it feeds. Pupation occurs within the webbed galleries.

Palm Leaf Skeletonizer = Order: Lepidoptera; Family: Momphidae

---

_____ **CONTROL CLUE** _____

Palms may be sprayed with Sevin 50% WP, or infected leaves and the interleafing husk may be removed and burned. Either method must be used persistently.

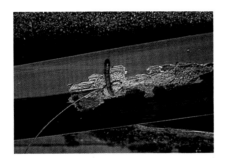

Palm Leaf Skeletonizer; caught in the act
(DPI)

Palm Leaf Skeletonizer frass on canary island date palm
(DOF)

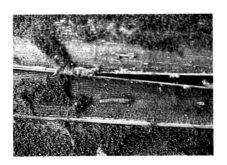

Palm Leaf Skeletonizer — the inside story
(DPI)

Pantry Pests

A variety of insects infest nearly any type of dried food stored in the pantry. These pests may be brought in as eggs or larvae in purchased food, or adult insects may be attracted to food for egg laying from nearby infestations within the home or from outdoors.

ANGOUMOIS GRAIN MOTHS are similar in size and color to clothes moths, but the larvae feed only on whole kernels of corn, wheat, popcorn, indian corn decorations, seeds in dry flower arrangements and bulk stored grains. No woolens. Adult moths will fly around homes in the daytime, while clothes moths shun the light. This is the tiny, fluttering critter that will watch television with you at night and cause your wife to run for the moth balls.

GRAIN BEETLES, FLOUR BEETLES, CIGARETTE BEETLES AND DRUGSTORE BEETLES are small, reddish-brown to cinnamon colored beetles that enjoy collecting around light fixtures. The larvae are small, cream-colored worms with dark brown heads. They infest flour, dried pet food, meal, macaroni, cereals, crackers, prepared cake mixes, spices and dried fruits. The adults feed on the same foods. (If you find some of those little brown-headed worms in your flour supply, don't waste the flour. Bake a fast chocolate cake. No one will ever know; I won't tell.)

DERMESTID BEETLES are scavengers on plant and animal products and will feast on leather, furs, skins, dried meat products, woolens and silk materials, cheese and cereal grain products. Dermestids can be divided into three categories based upon the type of food preferred:

LARDER BEETLES and larvae prefer products of animal origin and may infest dried meats and cheese. They are rarely found on foods of plant origin. You will more likely find these critters out in the compost box rather than in the pantry. CARPET BEETLES also prefer products of animal origin, but may be found throughout the home feeding on carpets, clothing, upholstery and wool or silk fabrics. They are occasionally found in stored food products. If you discover carpet beetles in your home, call a qualified pest control service. Carpet beetles can cause you great damage and cost you big money. CABINET BEETLES are the only group of dermestids that prefer cereals, grain products, spices and other true pantry items. The larvae do most of the damage.

FLOUR MOTHS are about ⅓ inch long. Typical examples are the INDIAN MEAL MOTH and the MEDITERRANEAN FLOUR MOTH. The larvae will leave their food site and wander around searching for a suitable place to pupate. The moths are strong fliers and often are found flying in proximity to the infestation site.

BROWN SPIDER BEETLES are about ¼ inch long and are equipped with long spindly legs. They rather look like spiders, but they aren't. Both adults and larvae are active feeders.

MEALWORMS, PSOCIDS and GRAIN MITES may be found where foods become moist or moldy. Destroy the infested food and correct the moisture problem.

Angoumois Grain Moths = Order: Lepidoptera; Family: Gelechiidae
Grain Beetles = Order: Coleoptera; Family: Cucujidae
Flour Beetle = Order: Coleoptera; Family: Tenebrionidae
Cigarette Beetle = Order: Coleoptera; Family: Anobiidae
Drugstore Beetle = Order: Coleoptera; Family: Anobiidae
Dermestid Beetle = Order:Coleoptera; Family: Dermestidae
Larder Beetle = Order: Coleoptera; Family: Dermestidae
Carpet Beetles = Order: Coleoptera; Family: Dermestidae
Cabinet Beetles = Order: Coleoptera; Family: Dermestidae
Indian Meal Moth = Order: Lepidoptera; Family: Pyralidae
Mediterranean Flour Moth = Order: Lepidoptera; Family: Pyralidae
Brown Spider Beetles = Order: Coleoptera; Family: Ptinidae
Mealworms = Order: Coleoptera; Family: Tenebrionidae
Psocids = Order: Psocoptera; Family: Psocidae
Grain Mites = Class: Arachinda; Order: Acari; Family: Tyroglyphidae
Granary Weevil = Order: Coleoptera; Family: Curculionidae
Rice Weevil = Order: Coleoptera; Family: Curculionidae
Cadelle Beetle = Order: Coleoptera; Family: Trogositidae
Lesser Grain Borer = Order: Coleoptera; Family: Bostrichidae
Saw-toothed Gr..n Beetle = Order: Coleoptera; Family: Cucujidae
Confused Flov er Beetle = Order: Coleoptera; Family: Tenebrionidae
Red Flower Beetle = Order: Coleoptera; Family: Tenebrionidae

_____ **CONTROL CLUE** _____

Simply put, seek out the source and destroy. Thoroughly wash food containers and shelves with hot, soapy water; store fresh pantry food products in your freezer; call your pest control operator for the tough control problems.

Angoumois Grain Moth adult
(USDA)

Angoumois Grain Moth larva
(USDA)

Flour Beetle adult
(USDA)

Saw-toothed Grain Beetle
(USDA)

Cigarette Beetle
(USDA)

Red Flour Beetle adult
Van Waters & Rogers
1983 division of Univar

Red Flour Beetle adult
(VWR)

Dermested Grain Beetle larva
(USDA)

Dermested Grain Beetle adult
(USDA)

CIGARETTE & DRUG-STORE BEETLE

GRID 2 MM

Cigarette Beetle and Drugstore Beetle
adult comparison
(VWR)

RED & CONFUSED FLOUR BEETLE

GRID 2 MM

Red Flour Beetle/Confused Flour
Beetle comparison
(VWR)

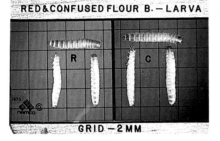

RED & CONFUSED FLOUR B. — LARVA

GRID — 2 MM

Red Flour Beetle/Confused Flour
Beetle larvae comparison
(VWR)

Indian Meal Moth — adult and larva
(USDA)

MEDITERR-ANEAN F. M. INDIAN M. M. ALMOND M.

GRID 5 MM

Indian Meal Moth/Mediterranean Flour
Moth/Almond Moth adult comparison
(VWR)

MOTH LARVA
INDIAN MEAL M.
MEDITERRANEAN FLOUR MOTH
ALMOND MOTH
GRID 5 MM

Indian Meal Moth/Mediterranean Flour
Moth/Almond Moth larvae comparison
(VWR)

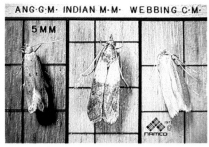

Indian Meal Moth/Angoumois Grain
Moth/Webbing Clothes Moth
comparison
(VWR)

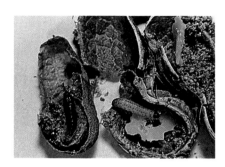

Almond Moth larva damage to peanuts
(USDA)

Psocid (book louse)
(VWR)

Brown Spider Beetle larva/pupa/adult
comparison
(VWR)

Cowpea Weevil larva inside blackeyed
bean
(VWR)

Cowpea Weevil Adults on blackeyed
bean
(VWR)

Mealworms — yellow and dark
(VWR)

Pecan Catocala

Several species of Catocalas may strip pecan leaves in the spring, leaving only the midribs. When fully grown, the caterpillars are dark gray and are about 3 inches long. They are loopers and become very active when disturbed. Both moths and caterpillars are so well camouflaged they blend with the bark when resting on trees and are frequently overlooked.

Pecan Catocala = Order: Lepidoptera; Family: Noctuidae

CONTROL CLUE

Although most catocala larvae reach maturity before time to spray for the pecan nut casebearer, most will be killed when that application is made. If perchance they get the jump on you and begin stripping too many leaves, don't wait. Hit 'em with Zolone. If you're lucky, the biggest problem you'll have with this insect is pronouncing its name . . . ka-TOCK-uh-la.

Pecan Catocala larva
(TAEX)

_____Pecan Cigar Casebearer_____

Pecan Cigar Casebearers may be more or less a problem, varying in incidence and consequent degree of damage from year to year. Mamas are moths. The feeding larvae produce tiny holes in pecan leaves. The most unique feature of this critter is its abode. It constructs a light brown, cigar-shaped case about a ¼ inch long in which it remains throughout development. The pecan cigar casebearer is not Pecan Public Enemy #1, but it is a most curious little creature that sometimes runs amok and requires specific control measures.

Pecan Cigar Casebearer = Order: Lepidoptera; Family: Coleophoridae

Pecan Cigar Casebearer on twig
(TAEX)

Pecan Cigar Casebearer and damage
(TAEX)

_____Pecan Nut Casebearer_____

The Pecan Nut Casebearer is perhaps the major pest of Florida
pecans. In early spring, the larvae of the overwintered generation feed
first in buds and then in developing shoots, causing both to wither and
die. Larvae of succeeding generations feed on nuts during late spring and
summer. Severe infestations may destroy your entire pecan crop. Are you
paying attention? Adult casebearers are light gray moths that are about a
⅓ of an inch long. They fly at night and hide during the daytime. Young
larvae (worms) are first white to pink in color, but later become olive gray
to green and grow to about ½ inch in length. First generation larvae hatch
from eggs in 4 or 5 days and migrate to buds below the nuts to feed. After
a day or two they enter the nuts and in feeding frequently destroy several
or all of the small pecans in a cluster. And if you don't do something early, you can kiss your pecan crop good-bye.

Pecan Nut Casebearer = Order: Lepidoptera; Family: Pyralidae

Pecan Nut Casebearer adult, pupa and
eggs
(TAEX)

Pecan Nut Casebearer damage in
terminal growth on pecan tree
(TAEX)

Pecan nuts damaged by casebearer
larvae
(TAEX)

Pecan Phylloxera

Pecan Phylloxera produces galls on new pecan growth. Leaves, twigs and nuts may be affected. Phylloxera passes the winter as eggs nestled in bark crevices. By spring, the tiny nymphs (babies) emerge and feed on tender young growth, secreting a substance that stimulates plant tissue to develop into galls. Adults are soft-bodied critters closely related to aphids. Upon maturing, phylloxera adults deposit numerous eggs inside these galls which split after 1 to 3 weeks, thereby liberating the next generation of nymphs to go after fresh, young growth on the tree. Several generations follow during summer and fall.

Incidentally, galls are conspicuous swellings which may be 1/10 to 1 inch in diameter. I understand some curious folks, presuming galls to be a fruit of the tree, have actually nibbled a few. (They taste lousy.)

Pecan Phylloxera = Order: Homoptera; Family: Phylloxeridae

_____ CONTROL CLUE _____

Dormant oil spray recommended for obscure scale will normally handle phylloxera. Apply thoroughly to tree trunks and limbs during the winter when trees are dormant.

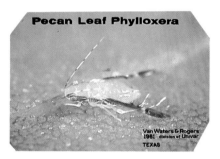

Pecan Leaf Phylloxera adult
(VWR)

Galls on pecan leaves caused by Pecan Leaf Phylloxera
(TAEX)

_____ Pickleworms & Melonworms _

Guess what Pickleworms like best to eat? You lose! Summer squash is tops on their menu, but cucumber and muskmelon are big favorites also. The larvae do the damage by eating on blossoms and vines and by invading the underside of fruits. Larvae (worms), bright green and dotted, grow to about ¾ inch in length and will likely be found pushing out small masses of green sawdust-like excrement (frass) from holes in the fruit they foul. These fruit soon rot, sour and mold after the interior has been exposed to air and microorganisms (bacteria and fungi). Fruits of late season plants are severely attacked.

Mama moths have dark brown wing margins which merge into lighter areas toward the center. The abdomen tip is tufted with hairs. Usually 4 to 5 generations are produced per season in Florida.

Melonworms have much in common with pickleworms. They eat at the same table, feasting on muskmelon, cucumbers and squash. Larvae are also bright green, but have dorsal white stripes running the length of the body; and they are larger, growing to an inch and a quarter in length. They feed on foliage rather than blossoms before tunneling into stems and fruit. Melonworm mamas (moths) have velvety-black wing margins with lighter, pearly-white areas.

Pickleworms and Melonworms = Order: Lepidoptera; Family: Pyralidae

_____ **CONTROL CLUE** _____

Control is difficult. The caterpillar must be killed before it bores into the fruit. Once entry has been made, it is impossible to reach these Dead End Kids with spray and the fruit is immediately ruined. Don't tarry. Liquid Sevin will do it to 'em. Begin treating at first sign of damage and repeat weekly till you clean them out.

Young larva of the Pickleworm
(USDA)

Older Pickleworm larva and damage
(USDA)

Pillbugs & Sowbugs

Pillbugs and Sowbugs normally live outdoors. They really are not insects, but are more closely related to shrimp, crabs and crayfish. Because they have difficulty in maintaining body moisture which is necessary for their survival, they habitually remain beneath objects on damp ground or even below ground during the day.

Both of these critters feed on decaying organic matter, but occasionally will feed on young plants or their roots. They are most attracted to over-ripe fruit lying on the ground. In my garden they give my strawberries fits. The compost pile is Thanksgiving dinner every day for these crustaceans; a citrus peeling or a rotten apple can be a gourmet treat.

Pillbugs and sowbugs maintain similar life-styles, but with one notable difference. When disturbed, pillbugs roll up into a ball; sowbugs do not. They sometimes enter houses, but do no damage.

Pillbugs and Sowbugs = Class: Crustacea; Order: Isopoda

_____ CONTROL CLUE _____

A pillbug and sowbug bait containing Sevin, Diazinon, Malathion or Methoxychlor is your best bet outdoors. Have no delusions about eliminating these dudes, for you're not likely to do that. Holding the population level down is really the most you can hope for. If they come in your house pick 'em up with the vacuum cleaner. Residual insecticides are usually not effective in the house.

PILL BUG & SOW BUG

Pillbug and Sowbug adult comparison
(VWR)

Praying Mantids

Here's a dapper fellow. When I was a kid back in Texas we called this critter a Grandfather's Walkingstick. Wrong name. Now, there is an insect called the Walkingstick; it is a large, usually wingless insect with legs all about the same length. Walkingsticks live on and feed on leaves of certain trees, occasionally causing damage. Praying mantids also have elongate, stick-like bodies. Those front legs fit together in order to hold prey.

Mantids range in size from ½ inch to over 6 inches and will feed on just about anything they can hold with their two front legs. They do no damage except to other insects, and that's why they are considered beneficial. They kill their prey by biting the back of the neck, which severs nerves and leaves the victim helpless. They have no qualms about feeding on one another. The larger feast on the smaller. After mating, the female may devour the male. (I've heard of a guy losing his head over a dame, but this is really going too far.) So, if you believe in reincarnation and intend to come back as a praying mantid, be sure you are big and are a female.

Many gardeners actually "sow" mantid eggs to encourage the presence of this critter, but mantids are relatively ineffective as pest controllers because they really aren't heavy feeders. The worst that can be said about the praying mantid is he is just as likely to prey on a harmless insect or even another beneficial insect, such as a honeybee, as he is on a pest. So, who bats a thousand?

The mantid hunts by waiting or by very slow stalking. When something appetizing passes too close, ZAP, and it's supper. No trouble in identifying the mantid. Nothing else looks like him. Body color usually is green, but some are brown or even pink. He can even turn his head to look around. Don't laugh . . . mantids need love too.

Praying Mantids = Order: Orthoptera; Family: Mantidae

Praying Mantid adult
(USDA)

Praying Mantids mating
(TAEX)

Psyllids

Psyllids are kin to aphids. The winged adults are small (1/16 inch) green or brownish critters that spring into flight with their large hind legs, prompting many to refer to them as "jumping plant lice." Psyllids damage plants by sucking plant juice and by transmitting disease; some cause blister-like galls to form on leaves. The wingless immature psyllids are often covered with white, waxy threads.

Psyllids feed on shoots and leaves of plants, causing distortion, stunting and often tip die-back. They may also cover a plant with honeydew, the undigested plant sap that is excreted as they feed. Psyllids show up on most all ornamentals at one time or another, but keep a wary eye on your persimmon and red bay.

When infestations are heavy, thousands of adults may swarm around and may even invade your home. They really won't hurt you, but will just "bug" you to death.

Psyllids = Order: Homoptera; Family: Psyllidae

Spray fruits and vegetables with an insecticide containing Malathion. Remember, read the label.

Common Willow Psyllid
(VWR)

Potato Psyllid adult
(VWR)

Red Bay Psyllids
(DOF)

Red Bay Psyllid damage
(DOF)

Potato Psyllid damage to tomato leaf; note leaf cupping
(VWR)

Rodents

The ability of Rats and Mice to survive in mankind's world is legend. They exist worldwide and will readily invade well-maintained suburban residences as well as garbage dumps. Evidence of a rat or mouse in your home should be cause for immediate action. The species that most frequently infest houses are the Norway Rat (also known as the house, wharf, brown or sewer rat) and the Roof Rat. Young rats can squeeze through an opening as small as ½ inch wide. Mice will enter openings slightly larger than ¼ inch wide. Rats will breed in any secluded location outdoors, including heavy vegetation such as ivy or juniper groundcovers. They may burrow in the ground or climb your pecan tree and eat their fill every night. They will invade your garden and munch on tomatoes or eggplant and have strawberries for dessert. If you tolerate their presence outdoors, it will be only a matter of time till you find them feasting on leftovers on your kitchen table. If allowed to proliferate, rats and mice can pose a serious health hazard as well as be a significant economic problem, in that they both contaminate food with fetal droppings and urine and destroy much property by gnawing. The long front teeth of rats grow constantly and to keep them worn down, rats will gnaw on most anything, including clothing, wood, electrical wires or furniture. Rats occasionally bite people, especially sleeping infants. A rodent bite is serious and should be treated by a physician. Although we normally don't think of Squirrels as being rodents, they are and they will do mischief similar to that of rats. Squirrels also will invade your home, but will more likely set up shop in the attic rather than the kitchen.

Rodents = Class: Mammalia
Norway Rat = Order: Rodentia; Family: Muridae
Roof Rat = Order: Rodentia; Family: Muridae
House Mouse = Order: Rodentia; Family: Muridae
Tree Squirrel = Order: Rodentia; Family: Sciuridae
Ground Squirrel = Order: Rodentia; Family: Sciuridae

CONTROL CLUE

Rat or mouse inside the house . . . try a trap and bait it with bacon or peanut butter. Rats or mice in out-buildings . . . offer dry cereal anticoagulant baits in ¼ or ½ pound packets. For small buildings with just a few rats, 2 pounds should be sufficient. For many rats, use 4 or 5 pounds. Use at least 3 pounds for a residence. The average amount for a farmstead is about 10 pounds. Mouseproof your house by sealing the cracks. Control of rodents is really not this simple. Why don't you call a qualified pest control service. Squirrels . . . a pellet gun and a keen eye, but shooting might be illegal in your community, so take care. (I have a great Squirrel Stew recipe. I got it from Grandma Zak.)

Norway Rat
(VWR)

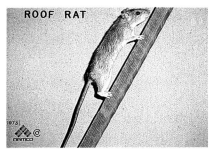

Roof Rat
(VWR)

House Mouse
(VWR)

Guess who came to dinner? Roof rats!
(VWR)

Tree Squirrel on ground
(VWR)

Ground Squirrels
(VWR)

Root Maggots

Not much we can say about Root Maggots except that we have them in Florida and they are a serious problem. Maggots are the larvae of certain flies and there are several species that prefer certain of our garden crops. There's the CABBAGE MAGGOT that primarily attacks cabbage, cauliflower, broccoli, radishes, beets and other similar crops. The ONION MAGGOT specializes in onions. The SEEDCORN MAGGOT will attack a variety of vegetable crops including corn, beans, peas, melons, cabbage, potatoes and turnips. Also included in the maggot passing parade — CARROT RUST FLY and SUGAR-BEET RUST FLY.

Maggots are usually white, sometimes bearing a yellowish cast. Larvae are about ¼ inch long when full grown. They work in the soil, chew off small roots and bore into the roots and underground stems of larger plants. Enough about maggots — they're no fun anyway.

Root Maggots = Order: Diptera; Family: Otitidae

CONTROL CLUE

Work some Diazinon granules into the soil and hope for the best.

Root Maggots on cabbage root
(USDA)

Rootworm, Southern Corn

We could have titled this segment, Cucumber Beetle Carnage, chapter 2! Remember, mama is the Spotted Cucumber Beetle. She's about ¼ inch long, greenish yellow in color and has 12 black spots on her back. The adults have a prominent head with relatively dark antennae. Adults are general foliage feeders, remember, and often occur in large enough numbers to cause considerable leaf damage. Damage occurs as irregular holes in leaves, damaged growing tips and occasionally the girdling of seedlings at or near ground level.

While the Spotted Cucumber Beetle takes the high road, baby (Southern Corn Rootworm) takes the low road. The major source of damage from this critter is brought about by the yellowish-white larvae (worms) that tunnel into and eat off the roots of plants. The lower part of the stalk is also often bored into. Once the plant is infested, little can be done. Southern corn rootworms thrive in wet soils that have a high clay content.

Although genetically different, very close kinfolks are the Northern Corn Rootworm and the Western Corn Rootworm. For your purpose, when you've seen one, you've seen 'em all. (Unless you're a rootworm.)

Southern Corn Rootworm = Order: Coleoptera; Family: Chrysomelidae

CONTROL CLUE

Control of rootworms is difficult and applications of insecticides must be made at planting time or shortly thereafter if this pest is suspect. A good shot would be Sevin or Diazinon for the old folks (adults) . . . Diazinon granules for the kids (rootworms).

Twelve-spotted Cucumber Beetle, adult
of Southern Corn Rootworm
(USDA)

Southern Corn Rootworm larva
(USDA)

Twelve-spotted Cucumber Beetle
(VWR)

____Rose Insects_____

The "superstar" of the flower garden is still the Rose. Many species of insects also find roses irresistible. Sucking insects insert their mouthparts into plant tissue and suck out the juices; chewing insects actually chew on plant tissue, thereby damaging all parts of the rose bush including roots, stems, leaves, buds and open blooms. Not all insects that frequent roses are damaging. Many are incidental; some are pollinators;

others are beneficial because they actually attack and feed on rose enemies. To control the enemies, rose growers must first identify them. Perhaps this page will be helpful.

Symptoms of sucking pests:
1. Wilting
2. Presence of honeydew
3. Curling and stunting of leaves
4. Yellowing of foliage
5. Dead spots in tissue

Symptoms of chewing pests:
1. Wilting (root damage)
2. Girdling of stem or cane
3. Severed stems, leaves or buds
4. Holes in leaves or stems
5. Discolored leaves or petals

ROSE APHIDS and other species relish roses. Species vary in color and may be black, green, yellow or pinkish. Watch for black, sooty mold on leaves. Plants appear wilted and sometimes leaves will yellow and drop, or curl and be stunted. Buds may fail to open, be deformed or produce small blossoms.

LEAFHOPPER adults and nymphs cause a stippling of the leaves. Adults vary in color from gray to yellow and green; some species have patterned markings. Nymphs resemble adults, but are lighter and are wingless. In Florida, leafhoppers may be found on roses from early spring until late fall.

Several SCALE INSECTS attack roses, but the most damaging is the soft Rose Scale. Female rose scales are round and dirty-white; males are elongate and snow-white. Mature scales insert their mouthparts into the plant tissue and remain fixed. They are most abundant with high humidity and reduced sunlight. They not only spoil the plant's appearance, but greatly reduce plant vigor.

Adult WHITEFLIES look like fluttering dandruff when disturbed. Immature whiteflies attach themselves to the underside of leaves and look much like scale insects. Both feed on roses and cause yellow spots on leaves. Heavy infestations can cause defoliation. Watch for that black, sooty mold.

TWO-SPOTTED SPIDER MITES are the most common of several species that attack roses. Their feeding-punctures appear as tiny light-colored spots, giving leaves a stippled appearance. Leaves of heavily infested plants turn yellow, then brown and eventually fall off.

FLOWER THRIPS, ONION THRIPS and TOBACCO THRIPS are the most common species that attack roses. Some are yellow to golden in color; others are almost black. Damaged tissue develops a silvery appearance; heavy infestations result in discoloration and deformed growth. Heavy feeding prevents buds from opening properly and results deformed, blemished petals.

Leaf-feeding beetles are a continuous problem for rose growers. Watch for: ROSE CHAFERS resemble June beetles and are brown in color. They can destroy a rose bloom in short order.

ROSE LEAF BEETLES are small metallic-green critters that feed in buds and on flowers. They riddle the blossoms with holes.

TWELVE-SPOTTED CUCUMBER BEETLES feed on tender rose leaves and buds.

LEAF and FLOWER FEEDING LARVAE are incidental pests, but most are voracious feeders. Only one or two can cause extensive damage. Good examples are the Corn Earworm and the Black Woolly Bear.

ROSE SLUGS are the immature stages of sawflies. Small larvae usually skeletonize leaves; larger larvae likely will eat the whole thing.

LEAFROLLERS are small, pale-green larvae with black heads that feed inside leaves that they have rolled up and tied with silk threads.

LEAF TYERS are similar to leafrollers in that they draw several leaves or part of leaves together and tie them with silk. They are light-green in color and are about ¾ inch long when full grown. They eat leaves.

ROSE STEM BORERS attack the stems or canes of roses. They are white larvae that bore into the sapwood and often will girdle the canes in several places. Some infested canes die back to the girdled area; others develop a swollen or enlarged area at the point of injury. Infested stems should be cut and destroyed.

LEAFCUTTING BEES, the phantoms of rose-land, are solitary bees that cut those circular pieces from rose leaves. The nature of damage is obvious.

GALL WASPS lay their eggs in rose stems. When the eggs hatch and young larvae begin to feed, the plant forms an enlarged area, or gall, in the damaged area. Infested tissue should be removed and destroyed before the wasps complete their development.

ROSE MIDGE is a serious problem for rose growers. The adult is a small yellowish fly. Female flies lay eggs in the growing tips of rose stems. The young larvae, or maggots, feed on the tender tissue, kill the tips and deform the buds. Remove and destroy infested tips daily to prevent maggot development.

GRASSHOPPERS feed not only on the leaves, but also on the buds, stems and blooms of roses. One or two grasshoppers can give your rose bush a rough time. Since grasshoppers migrate, control will be difficult.

Rose Aphids = Order: Homoptera; Family: Aphididae
Leafhoppers = Order: Homoptera; Family: Cicadellidae
Scale Insects = Order: Homoptera; Family: Coccidae, Diaspididae, Pseudococcidae
Whiteflies = Order: Homoptera; Family: Aleyrodidae
Two-Spotted Spider Mite = Class: Arachnida; Order: Acari; Family: Tetranychidae
Thrips (flower, onion and tobacco) = Order: Thysanoptera; Family: Thripidae
Rose Chafers = Order: Coleoptera; Family: Scarabaeidae
Twelve-spotted Cucumber Beetle = Order: Coleoptera; Family: Chrysomelidae
Rose Leaf Beetles = Order: Coleoptera; Family: Chrysomelidae
Rose Slugs = Order: Hymenoptera; Family: Tenthredinidae

Leafrollers and Leaf Tyers = Order: Lepidoptera; Family: Tortricidae
Rose Stem Girdler = Order: Coleoptera; Family: Buprestidae
Leafcutting Bees = Order: Hymenoptera; Family: Megachilidae
Gall Wasps = Order: Hymenoptera; Family: Cynipidae, Tenthredinidae
Rose Midge = Order: Diptera; Family: Cecidomyiidae
Grasshoppers = Order: Orthoptera; Family: Acrididae, Tettigoniidae

Rose Curculio adult and extensive
damage on rose bud close up
(VWR)

Rose Aphid winged migrant and
apterous female with pink nymphs
close up
(VWR)

Rose Aphid infestation on rose bud
(VWR)

Rose Curculio adults and damage to
rose petals of rose in bloom
(VWR)

A Rose Chafer
(VWR)

Western Rose Chafer adult on rose
(VWR)

Fuller's Rose Weevil and damage to eucalyptus leaf
(VWR)

Leafcutting Bee damage to rose leaves
(VWR)

Black Woolly Bear, just one of the wandering critters that will strip a rose bush in short order
(VWR)

Bristly Rose Slug mature larva, dorsal and coiled view on rose leaf
(VWR)

Bristly Rose Slug extensive damage to rose bush
(VWR)

Sawflies

Pine Sawflies can give Florida pine trees holy hell and they usually do. All southern pines of all ages are susceptible. Early damage is evidenced by reddish-brown straw-like remains of needles that are incompletely consumed by young larvae; older larvae leave only short stubs. Partially defoliated branches often have a "bottle brush" appearance. Occasionally, sawflies are a problem to ornamental trees.

Sawflies are not true flies; rather, they are wasps that cannot sting. They get their name from the saw-like structure on the end of the female's abdomen which she uses to lay her eggs. It is the larvae that consume needles or leaves and cause damage. Colonies may consist of just a few to over a hundred individuals, and they usually feed on a tree from top to bottom, completely defoliating one branch before moving to another. When disturbed, larvae may drop from branches, or assume a u-bend posture by raising their heads and abdomens. They can also orally exude a material that will paralyze insectan parasites or repel other predators.

Mature sawfly larvae spin cocoons in the duff or pine litter, mineral soil or under bark scales. Adult sawflies emerge by removing a cap at one end of the cocoon. After an appropriate courtship, marriage and mating, females lay eggs in slits sawed in pine needles. Fertilized eggs produce both male and female offspring, but the unfertilized eggs produce only females. (I wonder how they get away with that? This is Women's Lib taken a step too far.)

Full grown Red-headed Pine Sawfly larvae are about an inch long. They have a red-rust head and a hairless, yellowish-white body with 6 rows of black dots. Florida is home to several sawfly species: Redheaded pine sawfly; Blackheaded pine sawfly; Virginia pine sawfly; Slash pine sawfly; Abbot's sawfly and others. Markings will differ according to species.

Sand Pine Webspinning Sawfly = Order: Hymenoptera; Family: Pamphiliidae
Slash Pine, Redheaded, Blackheaded, Sand Pine, Spruce Pine, Abbot's, and a Pine Sawfly = Order: Hymenoptera; Family: Diprionidae

Suppression of sawfly populations by insecticides is usually successful. Sevin can do the job; spray foliage as soon as an infestation is observed. You may need to spray more than once. Read the label. Or you might try this old trick: gather sick larvae, grind them up, strain through cheesecloth and spray the juice on other sawflies. You may spread a viral disease that gut-rips sawflies. CAUTION: If you use the wife's blender to puree those larvae, you will be served divorce papers Monday morning.

Pine Sawfly adults (female and male) and cocoons
(DOF)

Redheaded Pine Sawfly larvae
(USDA)

Blackheaded Pine Sawfly larvae
(SFI)

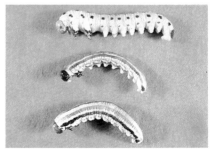

Redheaded Pine Sawfly (top);
Blackheaded Pine Sawfly (middle);
Slash Pine Sawfly (bottom)
(DOF)

Sawfly egg pockets in pine needle
(DOF)

Sawfly frass in ground litter
(DOF)

Oak Sawfly larva
(USDA)

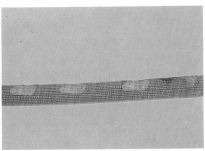

Slash pine defoliated by Sawflies
(DOF)

_____Scale Insects_____

Homeowners may have difficulty in controlling Scales because they tend to lose sight of the fact that scales are insects. Most mature female scales never move once they firmly attach themselves to branches, twigs or foliage; they let the males and nymphs (crawlers) do the walkin'.

There are hundreds of scale species found on a variety of host plants. A general differentiation is SOFT SCALES and ARMORED SCALES. Oval, soft-bodied Mealybugs, while not true scales, are closely related and will respond to scale controls.

Scales are weird lookin' critters which easily go unnoticed until both the infestation and consequent damage are extensive. At this point, the uninformed homeowner is likely to push the panic button and may go after this insect with a fungicide, or may apply the proper insecticide at the wrong time, or may apply the wrong insecticide at the right time. So, pay attention. Scales weaken or kill the host plant by sucking plant sap through piercing-sucking mouthparts. If scale populations increase, treat plants with insecticide oils during the dormant season, or treat with conventional sprays in the spring and summer. If you don't know what to do at a given time, ask a qualified person for advice.

Scale = Order: Homoptera; Family: Coccidae, Diaspididae, Asterolecaniidae, Kermidae

CONTROL CLUE

Use a dormant oil, or a conventional insecticide (Diazinon, Cygon or Orthene), or a contact insecticide (Pyrethrins), or a systemic insecticide (Di-Syston). The nature of the host plant will determine which control should be used. Remember, read the label and don't be afraid to ask. Successful control can be determined by sliding your thumbnail across a group of scales. If they are dry, hollow and flake off readily, they are dead. Some scales can be effectively controlled by daubing with rubbing alcohol or a mild soap solution. Scale is best controlled in the crawler stage.

Woolly Pine Scale on loblolly pine
(DOF)

Pine Scale
(SFI)

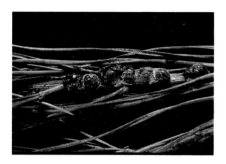

Pine Tortoise scale on pine needles
(DPI)

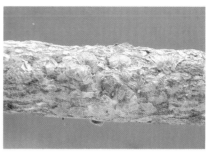

Holly Pit Scale on American holly
(DPI)

Obscure Scale on oak shoot
(DPI)

Oak Lecanium Scale on oak shoot
(DPI)

Florida Wax Scale
(DPI)

Kermes Scale on oak leaf
(DPI)

195

Latania Scale
(DPI)

Oleander Scale
(DPI)

Pyriform Scale
(DPI)

Dictyosperum Scale
(DPI)

Acuminate Scale
(DPI)

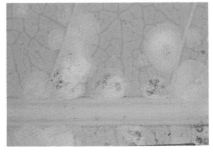

Mango Shield Scale
(DPI)

Stellate Scale
(DPI)

San Jose Scale
(DPI)

Hemispherical Scale and sooty mold
(DPI)

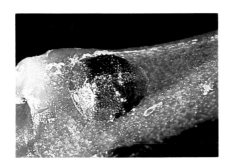

Mexican Black Scale
(DPI)

Red Wax Scale
(DPI)

Green Shield Scale
(DPI)

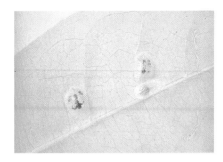

Indian Wax Scale
(DPI)

Brown Soft Scale
(DPI)

Terrapin Scale
(DPI)

Banana-shaped Scale
(DPI)

Tea Scale
(DPI)

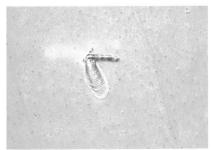

Purple Scale
(DPI)

Tuliptree Scale
(DPI)

Pine Tortoise Scale: leaf form
(DPI)

Striped Pine Scale
(DPI)

Nigra Scale
(DPI)

European Fruit Lecanium Scale
(DPI)

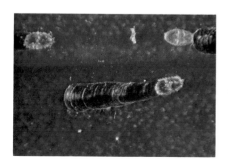

Black Thread Scale
(DPI)

Glover Scale
(DPI)

Citrus Snow Scale
(DPI)

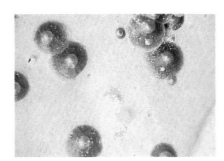

Florida Red Scale
(DPI)

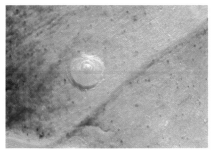

Yellow Scale
(DPI)

Cottony Cushion Scale
(DPI)

Pine Needle Scale
(SFI)

200

Barnacle Scale
(DPI)

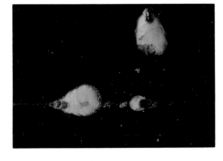

False Oleander Scale
(DPI)

Cottony Maple Scale
(DPI)

Camelia Scale
(DPI)

Duges Wax Scale
(DPI)

Green Scale
(DPI)

Japanese Wax Scale
(DPI)

Long Brown Scale
(DPI)

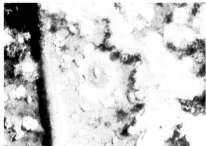

Coconut Scale
(IFAS)

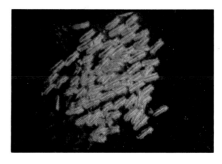

Lesser Snow Scale
(DPI)

White Peach Scale
(DPI)

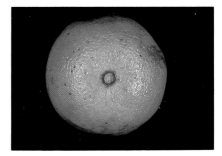

Chaff Scale
(DPI)

202

Scorpions

Scorpions occur throughout Florida and should be easily recognized by their large pincers near the head and by their thin tail carried over the back. They range in size from 1 to 5 inches, depending on the species, with colors ranging from yellowish-brown to black. The arched tail ends in a bulb-like poison gland equipped with a stinger. And that's what can hurt you. Forget those pincers. They are used only for holding food, which might be small insects, spiders, centipedes, other scorpions or earthworms. Scorpions hide under stones, bark of fallen trees, boards, firewood or other objects that lay on the ground.

Although individual reactions to the stings may vary, it is important to seek medical assistance immediately if a person, particularly a child, has severe reaction to a scorpion sting. Ice packs or alcohol swabs applied to the sting area are normally the suggested first-aid treatments.

Scorpions = Class: Arachnida; Order: Scorpionida

CONTROL CLUE

Chemical control for scorpions is not particularly effective. If you encounter one outdoors, hit it with a rock. If you see one indoors, step on it, but for God's sake, be sure you're wearing shoes!

A Florida Scorpion
(DOF)

Silverfish & Firebrats

Silverfish and firebrats are common house-invading pests. They are fast runners and are most active at night. They can go for long periods of time, sometimes over a year, without food, but will readily feast on flour, dried meat, rolled oats, paper and even glue. Considered mostly a nuisance, they damage paper goods, stain cloth and contaminate food.

Adults reach a length of about ¾ inch. Silverfish are covered with fine scales which are silvery to brown in color; firebrats are quite similar to silverfish, but are considerably darker in color. They are usually brought into the home in foodstuffs, furniture, old books and papers.

Silverfish and Firebrats = Order: Thysanura; Family: Lepismatidae

CONTROL CLUE

These critters prefer the dark, so give them light. Spraying with Dursban, Diazinon, Pyrethrins or Malathion should handle minor occurrences, but when infestations are large, persistent and hard to find, call your pest control operator. He has the knowledge, training and equipment to perform safe and effective control. Take my word for it.

Silverfish
(USDA)

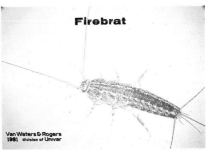

Firebrat

Van Waters & Rogers
1981   division of Univar

Typical Silverfish damage
(USDA)

Firebrat adult, close up
(VWR)

_____Snails & Slugs_____

Snails and Slugs are not insects — these critters are mollusks. They're related to clams, oysters and other shellfish and like the rest of the family, they must be moist all the time; consequently, they avoid direct sun and dry places. During the day, they hide under anything lying on the ground or they may secret themselves in ground-cover, weedy areas or compost piles. They emerge at night or on cloudy days to feed. In Spanish, slugs are called "babosas" from the verb "babear" (to drool), because of the copious amount of mucus they secrete. The mucus aids in their moving from place to place.

Snails and slugs are similar except that the snail has a hard shell into which it withdraws whenever it's pleased to do so. Snails and slugs often damage vegetables and garden crops. They feed on tender vegetation such as lettuce or Bird's-Nest Fern. Stems and leaves may be sheared off and eaten. Slugs are fond of mushrooms. Silvery trails winding around on plants and nearby soil is a definite clue that snails or slugs are present. Inspect the garden for them at night by flashlight.

Snails and Slugs = Phylum: Mollusca; Class: Megagastropoda; Order: Gastropoda; Family: Helcidae, Megalobulimidae, Bulimulidae

CONTROL CLUE

Clean surroundings will aid in controlling snail and slug populations. Removal of boards, sacks, piles of brush and other debris would be advisable. Natural enemies are relatively few, but some birds, especially ducks, feed on slugs. Snails and Slugs are attracted to beer. Put some out in a flat dish, like a pie pan, and in a day or two you'll have a dish full of dead critters . . . OR . . . You drink the beer and give the critters a Snail & Slug Bait containing Mesurol or Metaldehyde . . . or get a duck. I wonder which option you will choose?

Brown Garden Snails
(VWR)

Black Slug
(VWR)

Giant African Snail
(DPI)

_____Spiders _____

BENEFICIAL SPIDERS

Spiders enjoy about as much acceptance in society as did the Hunchback of Notre Dame. With only a few exceptions, these critters are not only harmless, but are beneficial. Although all spiders are capable of injecting venom when they bite, only a few, such as the Brown Recluse and the more infamous Black Widow are dangerous to people.

All spiders are beneficial predators on smaller creatures, and many are quite effective at reducing pests; many also employ some sort of webbing to do their thing, and people also tend to resent that web. Brother spider just can't win, can he! Let's take a look at some Florida residents.

The GREEN LYNX SPIDER is the spider most often received for identification by Division of Plant Industry entomologists. It is a conspicuous, large, bright-green spider found on many kinds of shrub-like plants. Although it is common throughout Florida and aggressively attacks its insect prey, it very seldom attacks humans.

Large ORB WEAVERS are common and conspicuous members in the moist woodland communities of Florida. They consume a variety of insects, and although typically nocturnal, adult females can be found sitting head down in the hubs of their webs in daylight hours.

The giant CRAB SPIDER, sometimes called the Huntsman Spider or the Banana Spider has sometimes been mistaken for a large Brown Recluse, but it is neither related nor is it dangerous.

One of three Jumping Spiders common in Florida is the REGAL JUMPING SPIDER which is capable of delivering a painful bite, but only if held tightly. The spiders are easily tamed and can be induced to jump back and forth from hand to hand. (This I gotta see!) They eat a bunch of insects and do not make webs.

The GOLDEN SILK SPIDER is a large orange and brown spider with feathery tufts on its legs. It is particularly despised by hikers and hunters who walk into its large, sticky web . . . usually face first.

The SOUTHERN HOUSE SPIDER is common throughout Florida. It's a large spider that makes a distinctive tangled, unkempt web. Males are frequently mistaken for the notorious brown recluse.

I really don't think you will be able to do this, but the next time you encounter one of these beneficial spiders . . . give the guy a break. Leave him or her alone.

Spiders (beneficial) = Class: Arachnida
Southern House Spider = Order: Araneida; Family: Filistatidae
Green Lynx Spider = Order: Araneida; Family: Oxyopidae
Orb Weaver Spiders = Order: Araneida; Family: Araneidae

Argiope Garden Spider = Order: Araneida; Family: Araneidae
Crab Spiders = Order: Araneida; Family: Thomisidae
Jumping Spiders = Order: Araneida; Family: Salticidae
Two-clawed Hunting Spider = Order: Araneida; Family: Clubionidae

_____ BENEFICIAL _____

Green Lynx Spider
(DOF)

Argiope Garden Spider
(DOF)

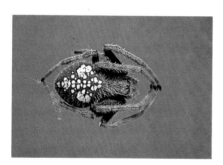

Orb Weaver Spider
(DPI)

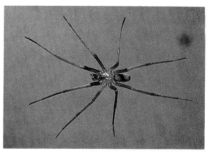

Southern House Spider — male
(DPI)

Southern House Spider — female
(DPI)

208

Yellow Crab Spider on a leaf
(VWR)

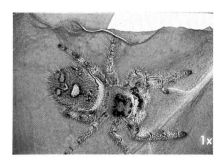

Regal Jumping Spider
(DPI)

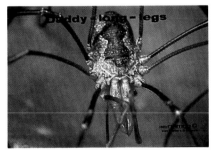

Daddy-long-legs Spider
(VWR)

Golden Silk Spider
(DPI)

Clubionid Hunting Spider
(VWR)

DANGEROUS SPIDERS

Of the nearly 3000 species of spiders that live in the United States, only two groups are considered dangerous to people and both are found in Florida. The better known are BLACK WIDOW SPIDERS; the lesser known are BROWN RECLUSE SPIDERS.

The BLACK WIDOW SPIDER is found outdoors in all kinds of protected cavities. Around homes she prefers garages, gas and electric meter boxes, furniture and many other unbothered places. A Black Widow bite feels like a pin prick and sometimes is not even felt. Usually, a slight local swelling and two red dots surrounded by local redness indicate the location of the bite. Pain becomes intense in 1 to 3 hours and may continue up to 48 hours. Symptoms include abdominal pains, a rise in blood pressure, nausea, profuse perspiration, leg cramps, tremors, loss of muscle tone and vomiting. The toxin also causes breathing difficulties and sometimes unconsciousness.

The BROWN RECLUSE SPIDER is a nonaggressive creature that spins a white or grayish, nondescript web. Its body and legs cover an area about the size of a quarter and its color varies from an orange-yellow to dark brown to almost black. The most distinguishing characteristics of this spider are its eyes and its back markings. It has three pairs of eyes arranged in a semicircle on the forepart of the head. The eyes also form the base of a violin-shaped marking on its back. The Brown Recluse often lives around human dwellings and is found in bathrooms, bedrooms, closets, as well as under furniture, behind baseboards and door facings or in corners and crevices. It also seems to prefer cluttered garages. Sometimes people are bitten while asleep; others may be bitten by spiders in stored clothing. Usually the bite causes a stinging sensation and then intense pain. Within 24 to 36 hours, a systemic reaction may occur, characterized by restlessness, fever, chills, nausea, weakness and joint pain. The bite also produces a small blister surrounded by a large conjested and swollen area. The venom usually kills the affected tissue, which gradually sloughs away and exposes underlying tissue. Healing may take 6 to 8 weeks, leaving scars that might require plastic surgery to repair.

If you are bitten by either of these spiders, seek medical aid immediately.

Spiders (dangerous) = Class: Arachnida
Black Widow Spider = Order: Araneida; Family: Oxyopidae
Brown Recluse Spider = Order: Araneida; Family: Loxoscelidae

_____ CONTROL CLUE _____

Know these spiders and their habits. Kill them on sight.

Black Widow Spider (underside view)
(USDA)

Brown Recluse Spider (note violin
marking at head)
(VWR)

Spittlebugs

Spittlebugs, also known as Froghoppers, are found in all geographic areas of Florida. Drops of undigested sap mixed with air are excreted by this insect, producing the frothy "spittle" that surrounds its body. This white froth is produced presumably to maintain an artificially high humidity which is required for development of the nymphs. The nymphs suck sap for a living and if infestations are severe, they can cause damage to host trees. On occasion infestations have damaged Pangola grass in the Everglades and Coastal Bermuda grass in northern Florida.

Adults resemble leafhoppers and fly actively during summer. They are yellow to grayish-brown in color, are wedge shaped and are about ½ inch long. Of course, the spittle-piles are the key to determining their presence on certain host grasses, pines, pecans and other trees. These critters act like they're always mad at somebody, but if I had to live in a pile of spit, I suppose I wouldn't be very happy either.

Spittlebugs = Order: Homoptera; Family: Cercopidae

Although these insects are becoming more common in Florida, they are not known to cause significant damage to pecan trees. Their incidence on pine trees, certain grasses and sometimes plants such as sugarcane and holly is cause for more concern. If they become a problem, spray with Isotox or Malathion. Read the label.

Australian Pine Spittlebug frothy mass
(DOF)

Pine Spittlebugs on pine
(USDA)

Spittlebug adult close up
(VWR)

Spittlebugs on pecan nuts
(USDA)

Squash Bug

When "looks" were passed out, this bug was behind the door. In addition to being the favorite contender-to-win in any UGLY contest, the Squash Bug sucks juices from leaves and stems of plants of the cucurbit family. They prefer squash, but will readily feast on pumpkins and melons.

Adults are brownish-gray to dark gray in color and range from ⅝ inch up to one inch in length. Nymphs usually have a green abdomen with crimson head, thorax, legs and antennae when first hatched, but later turn grayish-white with nearly black legs and antennae. You are likely to find a batch of these along with the adults.

Squash Bug feeding causes rapid wilting and leaves soon become blackened, crisp and dead. Some people think these critters are stinkbugs, because when crushed they emit a strong, disagreeable odor. Not so! I mean, they stink, but they aren't stink bugs.

Squash Bug = Order: Hemiptera; Family: Coreidae

CONTROL CLUE

Squash Bugs aren't simple to control. Hit 'em while they're little. Try Sevin. I have had the best luck with Ortho Tomato-Vegetable Dust which contains Methoxychlor and Rotenone. You can hand pick them, but don't do it with bare fingers, lest you become about as popular as a skunk-sprayed dog. That odor is tough stuff.

Squash Bug on leaf
(TAEX)

Stinkbugs

Stinkbugs are some of the most prevalent pests in Florida. Several species attack vegetable, field and fruit crops. They are sucking insects that prey upon beets, okra, squash, beans, peas, corn, tomatoes and many weeds. Damage is caused by nymphs and adults sucking sap from pods, buds, blossoms, leaves and seeds. If fruit is attacked at an early stage of development, "catfacing" (deformity) or pitted holes will occur on bean pods, tomatoes and squash. These feeding spots are surrounded by hard callouses, causing the fruit to become distorted and dimpled. If your tomatoes turn up with zits and your okra pods grow curl-y-cue like a pig's tail, most likely you have a case of stinkbugs.

The SOUTHERN GREEN STINKBUG is probably the most common of all the stinkbugs in Florida. It is a large flattened, shield-shaped, bright green bug about ⅔ of an inch long. The BROWN STINKBUG is similar to the green variety except that it reaches only ½ inch in length and is buff or brown in color. Normally the green outnumbers the brown, but there are times when they occur in reverse population numbers. Adults are winged and they do fly. Crush one and you will readily understand how the stinkbug got its name. Yuk! Hold your nose.

Some other members of this stinkin' family are the LEAFFOOTED PLANT BUG and the BIGLEGGED or BIGFOOTED PLANT BUG.

Southern Green and Brown Stinkbugs = Order: Hemiptera; Family: Pentatomidae
Leaffooted and Biglegged Plant Bugs = Order: Hemiptera; Family: Coreidae

CONTROL CLUE

In my vegetable garden, I have the best luck with a Pyrethrins spray. No waiting period after use. Wash your produce and eat it the same day. Read the label.

Brown Stinkbug
(USDA)

Southern Green Stinkbug
(USDA)

Several species of Stinkbugs that
attack vegetable gardens
(TAEX)

PREDACEOUS STINKBUG

Stinkbugs have a tarnished reputation as a group, but not every stinkbug is a bad guy. The predatory stinkbug has been collected throughout the year in Florida, but is more commonly found during the warmer months. It is most apt to be found on pole beans, potatoes, soy beans and other truck and row crops where its prey of beetle larvae and caterpillars abound. This gladiator is especially tough on larvae of the Mexican bean beetle, cabbage loopers and soybean loopers. Some specimens are solid iridescent blue, green or purple, while others are patterned with orange or red.

Take a look at Sir Lancelot squaring off with one of the villains. Here's a Predaceous Stinkbug attacking a green stinkbug. It is said, one picture is worth a thousand words. Need I say more about the predaceous stinkbug? If you're a bettin' man, let me give you this tip. Put your dough on the little red-and-black guy with the needle nose. I got it straight from the stinkbug's snout.

Predaceous Stinkbug = Order: Hemiptera; Family: Pentatomidae

_____ BENEFICIAL _____

Predaceous Stinkbug attacking another
stinkbug
(USDA)

Stinkbug predator "zonking" an
oleander caterpillar
(DOF)

_____Termites_____

Drywood (non-subterranean) and Subterranean Termites are the most destructive insect pests of wood, causing over $1 billion in damages each year in the United States; both termite types are prevalent in Florida. Termite food consists of cellulose obtained from wood. Protozoa in termite digestive tracts convert this cellulose into usable food. Drywood and subterranean termites require completely different control methods, so correct identification is imperative.

DRYWOOD TERMITES

Drywood Termites infest dry wood and do not require contact with the soil. Infestations may be found in almost any product containing cellulose and sometimes in animal products like hair or leather, but are most commonly found in building woodwork and in furniture. The surest evidence of infestation is the presence of fecal pellets. Drywoods spend their entire lives inside wood and seldom is there any visible evidence of their presence. They construct round "kick holes" in infested wood through which these fecal pellets are eliminated from the galleries or tunnels. These pellets accumulate in small piles below the kick holes or will be scattered if the drop distance from the kick holes to the floor is great enough. These pellets are distinctive and are used for identification. They are hard and elongated, are less than 1/25 inch long, have rounded ends with 6 flattened or concavely depressed sides with ridges at angles between the 6 surfaces. Wood damage, though seldom found, along with these pellets and perhaps a few discarded wings is rather conclusive evidence that . . . you got 'em.

Galleries or tunnels in the wood made by drywood termites cut across the grain, destroying both the softer spring growth and the harder summer growth of infested wood. At certain times of the year during evening hours, kings and queens emerge from the colonies in a swarm. In Florida, peak swarming periods occur January through June and again in October. There may be swarms to a lesser extent during other months. The soldiers and immatures remain in the wood at all times, and are seldom seen.

Drywood swarmers are commonly found in fence posts and pilings. They may enter a building through any opening. Garage interiors, particularly in unfinished garages, should be carefully observed since doors are left open for long periods. A sound coat of paint will deny access to a given surface, however, unpainted cracks or joints in that surface will still be subject to attack. No effective way has been developed to prevent entry through or under wood shingles.

Drywood Termites = Order: Isoptera; Family: Kalotermitidae

Fumigation is the most positive method of drywood termite control. First, a structure is completely enveloped in gas-proof, heavy nylon-plastic sheeting, then a toxic gas is released into the structure. The process is extremely hazardous and occupants must vacate the premises for at least a day. Yes, it's expensive, but you can't handle this chore by yourself. Call a pro. Select a certified pest control company that offers this service.

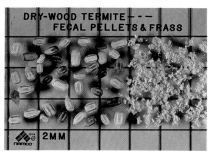

Drywood Termite fecal pellets and frass — symptoms of Drywood Termite activity
(BS)

Drywood Termite; alate (winged adult)
(BS)

Western Drywood Termite alates and workers
(BS)

Drywood Termite fecal pellets
(BS)

Example of a fumigation treatment commonly used for large-scale infestations of the Drywood Termite
(SFI)

SUBTERRANEAN TERMITES

Subterranean Termites nest in the soil. Moisture is important because they dehydrate readily. To survive, they must maintain contact with the soil (their primary moisture source) or other above-ground sources such as structures having defective plumbing or guttering. They must also protect themselves from temperature extremes and from attack by natural enemies such as ants and other insects; consequently, they build shelter tubes. These tubes are constructed by the worker termites from particles of soil or wood and bits of debris held together by salivary secretions.

Dead trees and brush are the original food source of subterranean termites. When land is cleared of this cellulosic material and houses are constructed on these sites, these structures become subject to attack. Termites can enter buildings through wood that is in direct contact with the soil, or by building those shelter tubes over or through foundations, or by entering directly through cracks or joints in and under foundations.

Generally, the first sign of infestation homeowners notice is the presence of swarming reproductives on window sills or near indoor lights. Other indications are the presence of wings, discarded by swarmers as a normal part of their behavior, or those shelter tubes going up the sides of piers, foundations or walls. Since our topic is subterranean termites, let me tell you about a real "lulu."

The FORMOSAN TERMITE was first identified in the United States through the efforts of my friend, Bill Spitz, Big State Pest Control, Houston, Texas. This termite was discovered infesting a dock-side warehouse at the Port of Houston in 1964 and since that time has been found in other states including Florida. In July, 1980 it was found causing damage to a Hallandale, Florida condominium.

Formosans swarm massively, usually between 7 and 10 p.m. This time of swarming is a good identifying symptom; however, a Florida dampwood termite and some drywood species are also evening swarmers.

Formosan termite nests are made of chewed wood, saliva and excrement and this material is called carton. No other termite in the United States constructs such nests. The primary nest is usually constructed near a source of food in or on the soil. Like other termites, the Formosans feed on cellulose and studies have shown that they can destroy wood 6 times faster than our native species. They have been known to penetrate lead, asphalt, plaster, mortar, creosote, rubber and plastic to get to wood underneath. They do this by secreting an acid substance from the frontal gland. Something else to worry about, they are known to attack 15 different kinds of living trees. These hombres are tough! Knowledgeable pest control operators ordinarily employ approved insecticides at double strength on the Formosan termite.

Powder-Post Termites, Florida Dampwood Termites and Southern Dampwood Termites are other species that torment Floridians; these pests have their own life-styles, but the result is the same . . . destroyed wood.

Subterranean Termite = Order: Isoptera; Family: Rhinotermitidae
Formosan Termite = Order: Isoptera; Family: Rhinotermitidae
Southern Dampwood Termite = Order: Isoptera; Family: Kalotermitidae

_____ CONTROL CLUE _____

This is not a do-it-yourself project. Termite treatment often requires specialized equipment such as drills, pressure injectors, pressure generating pumps, high-gallonage tanks and highly specialized knowledge. Superficial treatment is as bad as no treatment. Call a qualified, professional pest control service.

P.S. There ain't nothin' funny about termites!

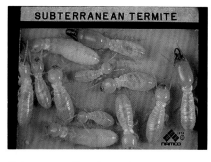

Subterranean Termite workers, soldiers and reproductives
(VWR)

Earthen shelter tubes of subterranean termites
(SFI)

Subterranean Termite workers
(SFI)

Termite alate; note dark body of this
native subterranean species
(SFI)

Alates of Formosan Subterranean
Termite
(SFI)

Formosan Termite damage to
heartwood of living ash tree
(SFI)

Carton nest of Formosan Termite used
to store moisture; this extremely hard
mass can be found crusted in wall voids
from floor to ceiling in infested
structures.
(SFI)

Mixed castes of the Formosan Termite
(SFI)

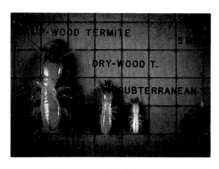

Drywood/Dampwood/Subterranean
Termite soldier comparison
(VWR)

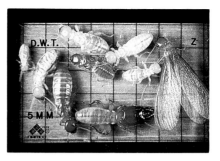

Dampwood Termites — workers,
soldiers and reproductives
(VWR)

_____Thorn Bug_____

The Cat's Claw Thorn Bug is an occasional pest of ornamentals and fruit trees in southern Florida. These treehopper kinfolks are generally not found north of Winter Haven or south of Florida City. Typically, the adult is about an inch in length and is green or yellow with reddish lines and brownish markings. Parasites and predators of the Thorn Bug are conspicuous by their absence . . . and no wonder. Check those hard, thorny spines on this guy's back. How would you like to eat a fish hook? There are reports of barefooted children stepping on these dudes. The wounds are slow healing and sometimes become infected.

Host plants that have been severely damaged by heavy infestations of Thorn Bugs include woman's tongue tree, acacia, hibiscus, bottle brush, dwarf date palm, citrus and others. Young jacaranda and royal poinciana trees have been killed by thorn bugs in the Tampa area. Honeydew secretions and accompanying sooty mold development, not uncommon where treehoppers prevail, can be a nuisance to homeowners.

Thorn Bug = Order: Homoptera; Family: Cercopidae

Experimental work on the control of this little monster is rather limited possibly because of its sporadic nature, but Diazinon, Cygon, Lindane or Malathion should do the job when control is called for.

Cat's Claw Thorn Bug
(DOF)

Thrips

Thrips are barely visible insects, less than 1/25 inch long. Many gardeners suffer thrip damage and never determine the cause. You can hardly see these darn things.

These tiny, slender, spindle-shaped rather active insects vary from pale yellow to yellowish-brown to black in color. Four slender wings are present on the females and these wings are fringed with comparatively long hairs on back margins. Males are wingless. Larvae resemble adults, but have no wings and are smaller. And, try to figure this one out . . .

223

adult females can reproduce regularly without mating with the rarely found males (somebody really got a lousy deal here). And you think you have it rough sometimes!

Thrips rasp the plant tissue and drain the exuding sap. This action causes stunted and deformed plants and when thrip populations are excessive, they can actually kill the plant. Thrips are general feeders, attacking vegetables, flowers and field crops. And if you have Cuban laurel, you have, will have or have had the Cuban Laurel Thrip. Wanna bet?

Thrips = Order: Thysanoptera; Family: Thripidae, Phaleothripidae

_____ **CONTROL CLUE** _____

A Malathion or Diazinon spray is good enough. Read the label.

This is a Thrip
(TAEX)

Greenhouse Thrips — adults, fecal
material and damage
(VWR)

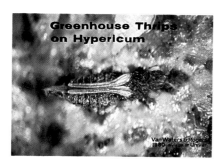

Greenhouse Thrip adult magnified
(VWR)

Greenhouse Thrips damage to
Hypericum groundcover
(VWR)

Thrips as you might see them on a white blossom
(ORTHO)

Cuban Laurel Thrips on Cuban Laurel
(DOF)

Cuban Laurel Thrips — nymph and adult
(VWR)

_____Ticks_____

Most of us, at one time or another, will experience the aggravation of tick bites. Knowledge of ticks and their habitats, as well as methods of tick control, can help you avoid this problem. Ticks are not insects, but are closely related to mites, spiders and scorpions. They are grouped into two families: HARD TICKS, which have a hard, smooth skin and an apparent head; and SOFT TICKS, which have a tough, leathery, pitted skin and no apparent head.

Hard Ticks usually mate on a host animal. The female then drops to the ground and deposits from 3,000 to 6,000 eggs which hatch into larvae or "seed ticks." Larvae climb nearby vegetation where they collect in large numbers while waiting for a host to pass within reach. After a blood meal on the host, the engorged larvae drop to the ground, shed their skins (molt) and emerge as nymphs. The nymphs locate a host, engorge themselves with blood, drop to the ground, molt and become adults and repeat the reproductive cycle. The term "wood ticks" is applied to several species of hard ticks so similar in appearance and habits that it is difficult to distinguish one from another. In Florida, the most commonplace ticks are probably American Dog Ticks and Brown Dog Ticks.

Adult AMERICAN DOG TICKS are chestnut-brown with white spots or streaks on their backs. Unfed adults are about ⅛-inch long; engorged females become slate-gray and may expand to a length of ½ inch. These ticks are widely distributed but are the most abundant in coastal or other humid areas. They are attracted by the scent of animals, so are most often encountered near roads, paths, trails and recreational areas. Although present the year round, American dog ticks are usually most numerous in the spring. Larvae and nymphs feed mostly on small rodents, while adults feed on dogs, cats, humans and other animals.

Adult BROWN DOG TICKS are reddish-brown. Unfed adults are ⅛ to 3/16-inch long; engorged females are about ½-inch long. They feed almost exclusively on dogs where they attach to the ears and between the toes. *They rarely attack man or other animals.* Inside the home, the ticks hide behind baseboards, window curtains, bookcases and cabinets, as well as inside upholstered furniture and under the edges of rugs. Outdoors, they hide near building foundations, in crevices between porch flooring, siding, and also beneath porches.

Because tick movements and bites are seldom felt, careful and frequent examination for ticks on the body and clothing is imperative. Early removal is important since many disease organisms are not transferred until the tick has fed two to eight hours. Always remove the tick with its mouthparts intact; hasty removal of an attached tick may break off the mouthparts and if they are left in the skin, they can transmit disease organisms or cause secondary infection. To relax tick mouthparts for easy removal, touch the tick with a hot needle or with a drop or two of camphor, alcohol, turpentine, kerosene or chloroform. Sometimes the best method is to grasp the tick firmly with tweezers or fingers and remove it with a slow, steady pull. Always treat the wound with a germicidal agent.

Ticks = Class: Arachnida; Order: Acari; Family: Ixodidae, Argasidae

American Dog Tick
(USDA)

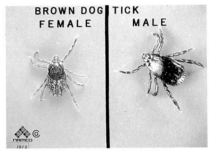

Brown Dog Ticks — female and male
(VWR)

____Tomato Pinworm_____

The Tomato Pinworm is a small, yellowish-gray or green, purple-
spotted caterpillar that can raise hob with tomatoes. In Florida, it is common in tomato producing areas south of Tampa along the west coast and
from Ft. Pierce south along the east coast. Mama is a moth. The tiny larva
does initial dirty work very much like that of a leafminer. It tunnels

through the leaf, eating away the chlorophyll tissue, leaving that crooked tunnel behind, but unlike the leafminer, the pinworm will come out and roll and tie leaf tips together. Another difference . . . better toilet training. It does not leave a trail of fecal matter in its tunnel like the leafminer, but returns to the entrance to make deposits. Big deal!

Tomato pinworm larvae are only about ¼ inch long when grown, but let me tell you what these squirts will do to your tomato patch. Initial injury is slight and appears as a small leaf mine. Later injury includes the leaf folding and leaf tying, but mature larvae may abandon the leaf and bore into the fruit leaving pin-size holes. Plant tissue becomes infected by pathogens and the fruit rots or the whole plant dies. Approximately 60 to 80 percent of your tomato crop may become infested in a single season. So, if you think you have leafminers that seemingly have stopped doodling and have begun housekeeping inside your tomatoes, your problem is likely not loco leafminers, but rather tomato pinworms.

Tomato Pinworm = Order: Lepidoptera; Family: Gelechiidae

_____ **CONTROL CLUE** _____

The Diazinon you apply to control the other tomato critters should likely handle this one also, but be sure you get him before he enters the tomato. Pay attention, or bye-bye tomatoes.

These tomatoes have been invaded by Tomato Pinworms. Pull 'em off the bush; even now they are useful only for chunkin' at the neighbor's dog.
(TAEX)

_____Treehoppers_____

Treehoppers are small, winged, sucking insects of peculiar and sometimes even bizarre shapes. They live on many plants, but because of their protective color and form, they are usually noticed only when moving. They are closely related to Leafhoppers and although they do suck sap, the primary injury to plants is the result of wounds made by the female; these are double rows of curved slits in the inner bark into which eggs are embedded.

Infested fruit trees, ornamentals and rose bushes look rough, scaly or cracked and seldom make vigorous growth. The fungi that causes rose canker and other diseases gain entrance to the plant through these bark slits.

Treehoppers = Order: Homoptera; Family: Cercopidae

_____ CONTROL CLUE _____

Treehopper eggs winter in the wood, so a dormant oil spray will be helpful here. In late spring, eggs hatch into spiny nymphs which drop from the tree and feed on sap of various weeds and grasses until mid-July or August when they become adults, so practice good housekeeping. Clean up! When adults are present on a plant, use an approved Malathion or Diazinon spray. Read the label.

Oak Treehopper female (stem mother)
(VWR)

Oak Treehopper male, a good-lookin' dude!
(VWR)

Oak Treehopper nymph, adult male and adult females
(VWR)

Oak Treehopper adult male, females, egg laying scars and feeding punctures
(VWR)

Buffalo Treehopper adult
(VWR)

Buffalo Treehopper nymph
(VWR)

Oak Treehopper in conference on oak shoot
(DOF)

Plant Hopper playing peek-a-boo on an azalea bush
(USDA)

Three-cornered Alfalfa Hopper — a country cousin visiting for the summer
(USDA)

Oak Treehopper damage to branch
(VWR)

Walkingstick

The Walkingstick is a defoliator of broadleaf trees. Earlier we mentioned this fellow as sometimes being confused with the praying mantid. These slender, wingless, stick-like insects are pale green when young, but gradually change to dark green, gray or brown at maturity. The adult female measures up to 3 inches in length and is more stout-bodied than the male.

Mating usually takes place in August; eggs are dropped to the ground where they overwinter in leaf-litter and normally hatch the summer after they are laid, usually starting in mid-May. The newly hatched walkingstick looks like a pint-sized adult.

Walkingsticks are rather docile in manner. If curiosity gets the best of you, and you're tempted to pick one up and bring it to your face for a closer look . . . DON'T DO IT. They emit a foul-smelling stuff that can be very painful if blown into your eyes. Tell the kids.

CONTROL CLUE

At times, walkingstick populations build in sufficient numbers to defoliate trees over large areas, but this is the exception, not the rule. Insecticide application is normally not practical. Parasitic wasps and flies are active against immature walkingsticks and perform a degree of population control; however, flocks of robins, blackbirds and grackles have a much greater impact. So, walkingsticks are strictly for the birds and let's just leave the matter there.

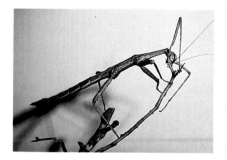

Walkingstick adult
(TFS)

Walkingstick hanky-panky; yes, he's small, but she likes younger men.
(DOF)

Wasps

PARASITIC WASPS

There are thousands of species of Parasitic Wasps. Adults are ordinarily minute or even microscopic in size, and that's the reason they pass unnoticed and mostly unappreciated. These darned half-pints are hard to see. Adults usually have black or metallic colors on the body.

Adult wasps lay eggs in or on the bodies of other insects which might be at various stages of development. After hatching, these wasp larvae feed inside or outside the body of their host. These parasites usually reduce the feeding of or vitality of the host and in due time will kill it. An excellent example of the activity is the tiny Braconid Wasp setting up housekeeping on a hornworm. So, it's a lousy lunch, but it's excellent pest control.

Braconid Wasp = Order: Hymenoptera; Family: Braconidae
Ichneumon = Order: Hymenoptera; Family: Ichneumonidae
Chalcids = Order: Hymenoptera; Family: Chalcidae
Prospaltella = Order: Hymenoptera; Family: Aphelinidae
Aphytis = Order: Hymenoptera; Family: Aphelinidae

_____ **BENEFICIAL** _____

Braconid Wasp parasite stinging an
oleander aphid
(VWR)

Braconid Wasp parasite with oleander
aphids
(VWR)

Braconid parasite cocoons on tomato
hornworm (Adios, hornworm!)
(DOF)

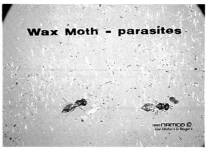

Ichneumon Wasp adult
(VWR)

Chalcid Wasp adults
(VWR)

PREDATORY WASPS

Predatory Wasps are present in a wide variety of colors, shapes and sizes. Solitary and colonial species occur. Adults can be ¼ inch or considerably more in length. Some are stocky in shape; others are svelte and thread-waisted. These are the wasps most encountered by people. In the process of stocking food for their larvae, just a few working predatory wasps can remove a significant number of insects from a garden. Most species are capable of stinging people and a few aggressively pursue that end if disturbed.

PAPER WASPS build unprotected paper nests out of chewed wood and hang them under eaves or in sheds or in bushes. Most of us call these wasps yellow jackets because they so much favor the true yellow jacket in appearance. True yellow jackets tend to live in much larger groups, usually building their nests underground.

Another wasp pest is the BALDFACED HORNET. This is a large (¾ inch long) blackish species with white markings. These hornets construct an inverted, pear-shaped, paper carton nest which can be up to 3 feet tall. These nests are usually built in trees, but may be attached to the side of buildings. Baldfaced hornet colonies are populated by hundreds of individuals which are very aggressive when aroused and their stings can be intensely painful.

Good examples of solitary predatory wasps are the king-sized CICADA KILLER, the MUD DAUBER and the POTTER WASP.

Paper Wasps = Order: Hymenoptera; Family: Vespidae
Yellowjackets = Order: Hymenoptera; Family: Vespidae
Hornets = Order: Hymenoptera; Family: Vespidae
Cicada Killers = Order: Hymenoptera; Family: Sphecidae
Potter Wasps = Order: Hymenoptera; Family: Vespidae
Mud Dauber = Order: Hymenoptera; Family: Sphecidae

CONTROL CLUE

Remember, these are primarily beneficial critters and should be eliminated only if their presence poses a threat to people. CICADA KILLERS: Control is rarely warranted. Males cannot sting; females will not sting unless forced to do so. MUD DAUBERS: Nests can be simply removed by hand with a putty knife, as females will not fight back. PAPER WASPS: Wait till dark; wasps will be docile and on the nest. By flashlight, soak the wasps and nest with one of the aerosol stream-sprays containing pyrethrins. When the wasps are dead, remove the nest. GROUND-NESTING YELLOWJACKETS: Locate the nest during the day, then return after dusk using a red light. Apply about one quart of Sevin spray directly into the nest entrance and then plug the hole with cotton. HORNETS: Nests that pose no threat to humans should be left unharmed, otherwise control and nest removal should be left to a professional.

Yellow Jacket Wasp — adult
(VWR)

Cicada Killer Wasp close up
(VWR)

Mud Dauber Wasp adult
(VWR)

Bald-faced Hornet — adult
(VWR)

Bald-faced Hornet nest in tree
(VWR)

Potter Wasp nest
(VWR)

_____Webworms_____

FALL WEBWORMS

Fall Webworms are frequently confused with the eastern tent cater-pillar. There are many similarities, but also there are many differences. The loosely woven, dirty-white webs produced by fall webworms are found on the terminal ends of branches. The larvae (worms) consume the foliage within the web. In Florida 3 or 4 generations occur per year.

Pecan trees are a preferred host, along with sweetgum, bald cypress, black walnut, persimmons, hickorys and other trees. Fall webworm larvae are pale yellow, spotted with black. They attain a length of one inch when fully grown and are covered with long black and white hairs. They have a voracious appetite, so don't ignore them.

CONTROL CLUE

If you regularly grow pecans successfully, you are spraying your trees at appointed times with an insecticide. This regular spraying should control fall webworms along with the other critters. Zolone is an excellent insecticide for this purpose. Dipel or Thuricide would be other choices. Webworms can also be burned if you can reach the web, but be careful with the fire. Tie a rag on the end of a cane pole with a piece of wire. Douse the rag with charcoal starter and light it. Briefly hold the fire under the web nest, and scorched worms will come tumbling down. If you're standing underneath, keep your mouth closed. Pruning the web-bearing limb is another option.

Fall Webworm adult and egg mass
(VWR)

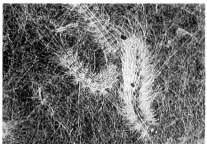

Mature caterpillars of Fall Webworm
(SFI)

Fall Webworms in nest
(VWR)

Fall Webworm caterpillars and damage
(SFI)

Fall Webworm nest
(USDA)

Heavy infestation of Fall Webworm on
persimmon tree in Marion County,
Florida
(DOF)

MIMOSA WEBWORM

The Mimosa Webworm is a destructive pest of ornamental mimosa. Larvae (worms) feed on the foliage of the host, often in semi-communal masses of silk webbing. Leaflets are tied together with the silk webbing by each larva. Sometimes the flowers are also attacked; young larvae actually prefer flowers over leaflets. Mature larvae drop to the ground via silken threads and seek protected pupation sites. There are several broods per year from May to September. Occasionally larvae become numerous enough to completely defoliate the host. A severe outbreak occurred in 1955 in Jacksonville.

The adult, a moth, is small, lustrous gray, with conspicuous black spots sparsely distributed on the forewings. Hindwings are uniformly unmarked dark gray.

The full-grown larva is gray to dark brown, with 5 white longitudinal stripes the length of its body. The dark background color may, on occasion, be diffused with pink and be interspersed with white markings which gives a mottled effect. The pupa is yellow-brown, enclosed within a silken, white cocoon.

Mimosa Webworm = Order: Lepidoptera; Family: Plutellidae

238

Mimosa Webworm adult moth
(DPI)

PINE WEBWORM

Pine Webworms derive their common name from the habit the larvae
have of chewing and gluing excreted pine needle material into a granular
nest. In Florida, pine webworms feed on loblolly pine, shortleaf pine,
slash pine, longleaf pine, Japanese pine, sand pine and Virginia pine.
Very young seedlings may die because of being completely defoliated. Infestations on ornamental pines around homes are unsightly.

Three to four generations develop per year. Adult parent moths appear in late spring and mate, but are seldom noticed because of their small
size (about 1 inch wingspan) and their non-descript coloring. When the
larvae hatch, groups of up to 75 wander among the needles spinning
silken threads. Each larva then bores into a needle and mines it.

Once they have grown too large to feed in individual needles, colonies of larvae feed among loosely webbed clumps of foliage, filling the
webbing with oblong, brown fecal pellets. It is an unsightly mess to say
the least. This mass of webbing and you-know-what may be 2 to 5 inches
long. Full grown larvae (worms) are about ¾ inch long, and are
yellowish-brown with dark brown longitudinal stripes on each side of
their bodies.

239

Often pine webworm larvae complete feeding and vacate their nests before the damage is noticed. It is too late to apply control measures if this has happened. If you find webworm nests that contain larvae (worms), prune the afflicted area and burn it. If pruning is not feasible, hit 'em with a strong shot of Sevin or Methoxychlor. A later application may be necessary.

Pine Webworm nest in
seedling top
(DOF)

Web and frass of the Pine
Webworm — a closer look!
(SFI)

SOD WEBWORM

The Tropical Sod Webworm is the #1 enemy of Florida lawns. Sod Webworms are the larval (worm) stage of lawn moths. Activity is most apparent during the early evening hours when moths can be seen fluttering above lawns as they drop eggs into the turf. Adults are small and vary in color from white to shades of gray. While at rest, adult moths hold their wings folded close to their bodies, making the snout-like projections on the fronts of their heads even more noticeable. The slender larvae may reach ¾ inch in length and are characterized by a light green to brown color with several rows of dark spots along the entire body.

During the summer months, sod webworm larvae live on the soil surface in silken tunnels constructed in the thatch of the grass. Damage occurs as larvae chew off grass blades and retreat into these tunnels to consume the foliage. Injury first appears as small brown patches of closely clipped grass. Lawns are particularly susceptible during the months of July and August when temperatures are hot and lawns are not growing vigorously. Large areas may be damaged rapidly if controls are not applied. Sod webworm larvae feed primarily at night and prefer areas in lawns that are hot and dry during daylight hours. Steep slopes, banks and other areas difficult to water properly are most subject to attack. Heavily shaded areas are seldom invaded.

Tropical Sod Webworm = Order: Lepidoptera; Family: Pyralidae

CONTROL CLUE

If 3 or 4 sod webworm larvae are found within a 6-square-inch section of dying sod, chemical treatment is recommended. Larvae are most active on cloudy days or at night. Prior to treatment, mow the lawn and rake infested areas to remove dead grass and plant debris. Use Dursban or Diazinon. If a granular form of either insecticide is your choice, a light watering after application will aid in achieving control; if you prefer an emulsifiable concentrate or soluble powder form, water the lawn thoroughly before application.

Tropical Sod Webworm larva in sod
(DPI)

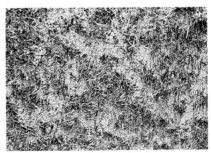

Tropical Sod Webworm damage to lawn
(DOW)

VEGETABLE GARDEN WEBWORMS

Several brands of Webworms give Florida vegetable gardens fits —
let's take a look at two of them. Garden Webworms and Beet Webworms.
Both are caterpillars and the parents are moths.

GARDEN WEBWORM adults are colored buff with shadings and
have irregular light and dark gray markings. Generally they are active at
night and are attracted to lights. Larvae (worms) are about an inch long,
are yellowish or greenish in color with a light stripe down the back. They
sport three dark spots on the side of each segment and these dots form a
triangle. Larvae feed primarily on the underside of leaves, skeletonizing
them. They spin webs and draw additional leaves into the web as more
food is needed. They are general feeders, attacking principally beans and
peas.

BEET WEBWORM adults are brown moths mottled with lighter and
darker spots. Larvae are about 2 inches long, are slender and are colored
yellowish to green with a dorsal black stripe. They also web leaves and
devour foliage, often migrate like armyworms and leave behind stripped
crops. Beet webworms attack beets, cabbage, beans, peas, carrots, spinach
and other crops.

Vegetable Garden Webworms = Order: Lepidoptera; Family: Pyralidae

Garden Webworm sitting for portrait
(TAEX)

Garden Webworm on vegetable plant
(TAEX)

Garden Webworms may be controlled with any one of a number of insecticides ... Dipel or Sevin or Diazinon. Both dusts and sprays are effective. These insecticides will be effective on any of the various species of webworms that are likely to harass you. You know, if you kill one ... you can kill 'em all.

_____Weevils _____

APOPKA WEEVIL

The common name for this weevil and its relatives in the West Indies is "fiddler beetle"; in Puerto Rico it is called the "sugar-cane root-stalk borer weevil" or "vaquita." It first appeared in Florida back in 1964 in a citrus nursery at Apopka, so we call it the Apopka Weevil.

Adults feed on foliage of a variety of plants, but most of the damage is confined to tender new growth. Citrus nursery trees often are affected more severely than mature grove trees. Feeding activity occurs both day and night. Adults are somewhat social in habits; often one tree may harbor hundreds while a nearby tree has none.

Mating takes place on the foliage. At Apopka, most specimens collected on the morning of October 3, 1968 were in copulation. (Tsk-tsk-tsk!) Eggs are laid in clusters between leaves stuck together with an adhesive produced by the female; she might lay 5000 eggs during her lifetime. The newly hatched grubs move across leaves in a peculiar galloping motion and fall off the leaf margin. They normally do not burrow into the ground immediately, but continue to move over the soil surface for several days. Eventually they burrow into the ground and find suitable roots for food. The most important hosts are sugarcane and citrus. Larvae are grubs and cause damage to small citrus trees by girdling the roots; this often kills the tree.

Apopka Weevil = Order: Coleoptera; Family: Curculionidae

243

Biological control seems to be successful in Puerto Rico. Experiments are continuing at Apopka to evaluate several chemicals under Florida conditions. Find something else to worry about, but keep an eye open for these critters. And if one morning you find a bunch of males and females practicing their naughty habit out on your orange tree, go in the house and call your local county agent. Don't stand there watching too long; that would be "kinky" on your part.

Apopka Weevil adult
(DOW)

CITRUS ROOT WEEVILS

A number of weevils are found on Florida citrus. Probably the most important native pests among these are Citrus Root Weevils. These weevils are often responsible for severe damage and sometimes death of citrus trees when the cambium of the tap root is completely girdled. Adult species of both Florida species have been found in every month of the year.

Adults are large, blue-green to gray weevils with relatively short, broad snouts. Eggs are laid in groups between two leaves or in folds of a single leaf. They hatch in about a week, then drop to the ground and burrow through the soil to suitable plant roots. Large roots are sometimes girdled and small ones often are totally consumed.

Citrus root weevils are commonly seen both day and night, especially on citrus. Only rarely have they been attracted to lights. Tree foliage is often heavily notched, but this damage is similar to that of all weevils known to feed on citrus.

Citrus Root Weevils = Order: Coleoptera; Family: Curculionidae

_____ **CONTROL CLUE** _____

Since these weevils often drop when disturbed, and their color contrasts sharply with the soil and debris under trees, they can be collected by shaking or beating branches. Why not!

Citrus Root Weevil
(DOW)

GARDEN WEEVILS

Weevils are a big family, in fact, one of the biggest as far as beetles are concerned. They are small, beetle-like critters with mouth parts modified into a downward-curving beak or snout. Weevils attack potatoes, tomatoes, turnips, carrots and many other vegetable, nut, fruit, grain and field crops. Several snout beetle varieties do dirty-work.

VEGETABLE WEEVILS are a variety that go after many Florida garden crops. Both larvae (grubs) and adults feed on the plants, primarily at night. Damage may resemble that of cutworms.

PEPPER WEEVILS are often a major pest where bell, chile and various other sweet peppers are grown. Eggplants occasionally serve as hosts, especially when grown in proximity to infested pepper fields. Adults are black, about ⅛ inch long with a sparse covering of tan to gray hairs. Young larvae tunnel into the seed mass in the center of the pepper pods. Pepper weevils have been found scattered over most of the Florida peninsula.

CARROT WEEVILS attack not only carrots, but also parsley and dill. Several generations occur in a single season and are a major pest. Grubs feed on the exterior or burrow into carrots with damage more likely occurring near the top. After weevils get done with a carrot, even Bugs Bunny wouldn't have it.

Vegetable, Pepper and Carrot Weevils = Order: Coleoptera; Family: Curculionidae

_____ **CONTROL CLUE** _____

Insecticides must be applied to kill the adult weevils before they lay eggs. Diazinon, applied thoroughly, will do a good job on brother weevil.

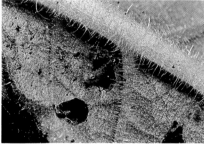

Vegetable Weevil larva and damage
(USDA)

Vegetable Weevil adult
(USDA)

Acorn Weevil close up; check that
schnozola!
(SFI)

Vegetable Weevil adult close up
(VWR)

GIANT PALM WEEVIL

The Giant Palm Weevil ain't kidding . . . it is the largest weevil in the United States. Length will vary from ¾ to 1 3/16 inches. Color is variable from uniformly black to mostly red with a variable black pattern. The surface is sometimes velvety, but usually somewhat shiny. The male's snout is covered with wart-like granules; the female's snout is smooth except for a few basal punctures. (So in the dark, they can tell who is who.) The larva is yellowish with an enlarged, dark brown head.

This weevil is known to breed in a variety of palms, but the adults also feed on several other fermenting plant products. They will attack healthy trees, but are most often attracted to weakened or dying trees, especially cabbage palms from which the "heart" has been removed for salad. Adults have been found during every month of the year in Florida.

The eggs are normally deposited in the bud of palms, especially cabbage palmetto and coconut palm. The larvae bore into the bud or leaf sheaths. They can cause the drop of large palm fronds or even the death of the tree. One species transmits a nematode which is the casual organism of "red-ring" disease of coconuts.

247

When feeding, the larvae make a most distinctive noise which has been described as being ". . . like the escape of water with an occasional screech like a choked hen." In the West Indies, larvae of one species are called "gru-gru" and are fried or roasted and are considered a delicacy. Also they are reputed to have the ability to produce milk in women, and the larvae of a European species is claimed to be a remedy for toothache.

Step right up, folks . . . get your hot "gru-gru" right here!

Giant Palm Weevil = Order: Coleoptera; Family: Curculionidae

CONTROL CLUE

Unless you need milk for the baby or have a toothache, I suppose you want to get rid of these giants. The older recommended control was to use a trap tree (by cutting the bud, then hand collecting the weevils). Adults can now be killed with several insecticides, one of which is BHC. And here's a good tip . . . palm trees with even slight injuries or bruises should be dusted with insecticide as a preventive measure. Remember that old saying . . . "An ounce of insecticide is worth several pounds of Giant Palm Weevils."

Giant Palm Weevil adult
(DOF)

Giant Palm Weevil feeding damage to a palm frond base
(DOF)

LEAF NOTCHER WEEVIL

The Little Leaf Notcher (LLN) is likely the most common weevil found in Florida's commercial citrus groves. Adults are easily seen on foliage because of their gray-white color; however, those from southern Florida often have a pink, green or even metallic gold cast. These "spooks" are closely related to the White Fringed Weevil and the Fuller's Rose Beetle. They have never been observed to fly and are probably flightless; therefore, they must either crawl or be carried from one location to another.

Although LLN shows a definite preference for citrus, it is nearly omnivorous on all groups of plants. Adults feed on foliage leaving a characteristic "notching" along the margin. Hatching larvae drop to the ground, burrow into the soil and feed on plant roots.

Adult weevils mate on the foliage of the host plant and females lay eggs in masses between closely touching leaves, gluing them together with a sticky substance. A mama may live for more than six months and lay over a thousand eggs in her lifetime.

Leaf Notcher Weevil = Order: Coleoptera; Family: Curculionidae

---

CONTROL CLUE

Although the Little Leaf Notcher may be locally abundant on some individual trees, it rarely is severe enough to require control . . . and that's a good lick for our side. Aldrin and Heptachlor are apparently effective controls, but neither is registered by EPA for use on citrus. Also, a number of natural enemies do a pretty good job of holding populations in check.

Leaf Notcher Weevil or *artipus floridanus* if you want to be a "show-off."
(DOW)

Fuller's Rose Beetle, "Notcher" kinfolks (Brother-in-law, I think!)
(DOW)

PECAN WEEVIL

Pecan Weevils are about ⅜ inch long and are brown. Adult females lay eggs directly in a pecan after making a hole with that long, efficient snout. Grubs that hatch develop rapidly as they feed inside the nut. They are creamy-white with reddish-brown heads. When the nuts drop to the ground in the fall, the developing grubs gnaw ⅛ inch holes through the shell. The grubs crawl through the holes and burrow into the ground.

It takes two or three years before adult weevils emerge again and attack nuts. In other words, this year's weevils will produce the ones you see two years from now. Pecan weevils are late season pests in several areas of Florida. In years when severe infestations occur, this son-of-a-gun may destroy a major portion of the area pecan crop.

Pecan Weevil = Order: Coleoptera; Family: Curculionidae

CONTROL CLUE

Begin checking trees in the first week of August to determine the presence of weevils. Weevil feeding may cause premature nut drop during the water stage of nut development. Homeowners who regularly spray their pecan trees with insecticide will also control the pecan weevil. Zolone does a good job for me.

Pecan Weevil larva
(USDA)

Pecan Weevil adult
(USDA)

REPRODUCTION & DEODAR WEEVILS

Adults of the PALES WEEVIL and the PITCH-EATING WEEVIL are attracted to stumps and dying trees where they deposit their eggs, but they also feed upon and destroy seedlings in the area. A weevil that is very similar, the DEODAR WEEVIL, normally breeds in the main stem of dying pines and in the past has been of little concern. In recent years, however, this weevil has been found infesting leaders of apparently healthy saplings and also it is disposed to attack and destroy seedlings weakened by almost any cause. These weevils are formidable pests in the areas of Florida where pine trees grow.

The Pales Weevil and Pitch-eating Weevil look much alike; adults are black or nearly so and are often speckled with whitish markings. They are ¼ to ⅓ inch long. The Deodar Weevil is slightly less than ¼ inch long and is a rusty-red with two distinct whitish blotches on the back. Females of the three species place their eggs in small holes in the bark of stumps and dying trees. The larva are small, white, legless and have shiny brown heads. They make separate winding tunnels between the inner bark and sapwood.

When fully grown, the Deodar larva constructs a conspicuous, oval "chip cocoon," ⅓ inch long, of finely shredded sapwood. Here it pupates and then, as a young adult, chews a round, BB-size exit hole through the cocoon and bark.

Most damage by the Pales and Pitch-eating Weevils occurs in spring and fall. The adults are attracted to the odor of freshly cut pine stumps, scorched pines or dying trees and they feed on the tender bark of nearby seedlings, causing significant damage. The Deodar Weevil breeds beneath the bark of saplings and larger pines that are dying from competition, suppression or bark beetle attack, but on occasion it attacks young seedlings which may be killed by the larvae tunneling in the cambium and woody tissue, or by multiple bark punctures caused by adult feeding.

NOTE: The Deodar Weevil is a vector of pitch canker fungus which can kill southern pines.

Deodar, Pales and Pitch-eating Weevils = Order: Coleoptera; Family: Curculionidae

_____ **CONTROL CLUE** _____

Prompt removal and disposal of pine stumps and of suppressed and dying pine trees is the best control for homeowners. You probably would be inclined to perform this "housekeeping" chore anyway, weevils notwithstanding.

Comparison of adults: Pitch-eating
Weevil (larger); Pales Weevil (middle);
Pissodes (smaller)
(SFI)

Pales Weevils — adults
(SFI)

Pales Weevil on pine
(USDA)

Pales Weevil larva
(SFI)

Pitch-eating Weevils — adults
(SFI)

Pine regeneration killed by Pissodes
Weevils
(SFI)

252

Deodar Weevils, pre-copulation on slash pine billet
(SFI)

Deodar Weevil chip cocoon
(DOF)

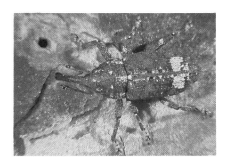

Deodar Weevil and feeding pit on slash pine
(SFI)

Deodar Weevil larva
(USDA)

_____Sweet Potato Weevil_____

The Sweet Potato Weevil is a problem in every county in Florida. It is the worst pest of sweet potatoes wherever it is found. This critter is worse than an Internal Revenue Agent for being persistent. This ugly devil never quits. Both adults and larvae damage the plants. Adults not only will feed on leaves and vines, but will also feed on and even breed in stored sweet potatoes. In the garden, adults lay their eggs on the plant near the soil surface. The grubs that hatch burrow into the vines and tunnel into the roots. There may be as many as eight generations in a single year, and they *don't* hibernate for the winter. They will subsist on weeds, especially wild morning glory, through winter. Seaside morning glory seems to be the favorite wild host. This weevil may cause almost complete destruction of a sweet potato crop.

Adults are about ¼ inch long and are ant-like in appearance. Their head, snout and wings are a dark metallic blue and their thorax and legs are bright orange. Another distinctive characteristic is that very long snout. Larvae are legless, brown-headed, white grubs almost ½ inch long when full grown. Invading sweet potato larvae impart a bitter taste to the tubers, often making them unfit for human consumption or even for stock feed.

Sweet Potato Weevil = Order: Coleoptera; Family: Curculionidae

_____ **CONTROL CLUE** _____

Don't plant sweet potatoes for a couple of years, and light a candle. Seriously, there are no chemical controls for this pest. Sorry!

Sweet Potato Weevil
(USDA)

Wheel Bug

The Wheel Bug is a fairly common, widely distributed, beneficial insect that preys on pest insects throughout Florida. It is one of the Assassin Bugs. Ordinarily there is only one generation a year, but in Florida some adults live into the winter months, particularly in the central and southern portions of the state. Some of the favored habitats of wheel bugs include cotton, goldenrods, sunflowers, trunks of locust trees, citrus and pecan groves, forest trees, shade and fruit trees. They prey on other insects and will jump anything from an aphid to a honey bee. Wheel bug saliva contains a toxic substance that immobilizes and kills its victims usually within 15 to 30 seconds after injection.

The adult wheel bug is ugly! It is a dark-complected, robust, grotesque creature with long legs and antennae. It is endowed with a stout beak, large eyes on a slim head and usually measures an inch or a bit more in length. Now, friend, that's what *ugly* is. The wheel bug is so named because of a distinguishing feature that is located just behind the head ... a prominent semicircular crest suggesting a cogwheel or chicken's comb. This is the only insect in the United States with such a crest. Now the bad news.

When captured, wheel bugs extrude (with little provocation) a pair of bright, orange-red scent sacs that give off a pungent odor. When disturbed, the wheel bug can inflict a painful bite that has been described as being similar to getting a typhoid shot with a large, dull needle; exaggeration notwithstanding, it's worse than stings from bees, wasps or hornets. The afflicted area often becomes reddish and hot to the touch, but later may become white and hardened at the puncture area. Initial pain is often followed by numbness for several days. These guys make "chirping" sounds, but don't be fooled. You ain't dealing with Tweety Bird ... wheel bugs are beneficial, but they *bite*.

Wheel Bug = Order: Hemiptera; Family: Reduviidae

_____ **BENEFICIAL** _____

(Unless you've been bitten by one.)

Wheel Bug
(SFI)

Whiteflies

Whiteflies are very tiny, snow-white insects that resemble moths if viewed under a magnifying glass. If viewed without magnification, they look more like flying dandruff. Adults are about 1/16 inch in length, have 4 wings and are covered with a white waxy powder. They resemble tiny white moths, but they are not moths; whiteflies are more related to scale insects. Nymphs (babies) are light green, oval, flattened and are about the size of a pin head. Their bodies are covered with radiating, long filament-like threads. You could confuse them with soft scale insects. Both nymphs and adults feed by sucking plant juices. Heavy feeding can give plants a mottled look, can cause yellowing and even death.

There are three major generations or broods of whiteflies in Florida. In the Gainesville area, these occur in late March, mid-June and late August. They will be approximately 2 to 3 weeks earlier in south Florida and about a week later in north Florida. Sticky honeydew excreted by these critters glazes upper and lower leaf surfaces permitting development of black sooty mold fungus; besides being unattractive, sooty mold interferes with photosynthesis, retards growth and also causes leaf drop. Whiteflies are common pests on many plants; some frequently attacked are allamanda, chinaberry, citrus, fringe tree, gardenia, ligustrum, viburnum, persimmon and many annuals.

_____ **CONTROL CLUE** _____

Malathion or Diazinon properly applied will ordinarily prevent ol' whitefly from getting a toe-hold, but if he gets a head start on you, spray twice or more per week. Cygon or Di-Syston applied as a soil treatment will do a good job as a systemic control if you apply it early enough. Remember, you will likely see flying "dandruff" first and later the black sooty mold.

NOTE: A tiny parasitic wasp, Prospaltella lahorensis, has done yeoman service in parasitizing citrus whitefly nymphs; the citrus blackfly is considered under complete biological control because of two tiny introduced wasps. 'Way to go, wasps!

Whitefly adults — side, dorsal and ventral views close up
(VWR)

Greenhouse Whitefly
(VWR)

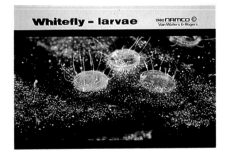

Whitefly larvae magnified
(VWR)

Citrus Whitefly adult close up
(VWR)

Black Sooty Mold on leaves
(VWR)

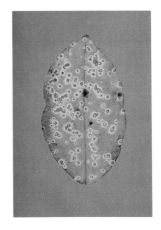

Whitefly fungus as it might
appear on a citrus leaf; I agree
it looks like the "crud," but
remember it is beneficial.
(DOF)

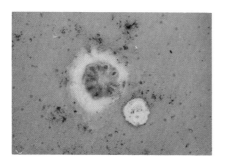

Whitefly fungus. Don't confuse this with
scale and eliminate it; this parasitic
fungus kills immature whiteflies.
(DOW)

White Grubs

This whole family is trouble. Adults are called May or June beetles; larvae are called Grub Worms. Adults will chew up leaves; grubs will feed on roots and other underground plant parts of many vegetable and garden crops. As many as 100 species may cause damage to vegetables.

People generally don't like these critters because they are so creepy. They really won't hurt you, but can cause you to hurt yourself. Hold an adult in your hand and it feels like he's eating to the bone (what you really feel is not teeth chewing but feet scratching). Let one sail into your wife's new hairdo (the adults do fly) and at the very least it will mean another trip to the beauty shop or maybe to the therapist, mental or physical.

White Grubs = Order: Coleoptera; Family; Scarabaeidae

_____ **CONTROL CLUE** _____

Chickens eat 'em, but who has chickens anymore! I've found that a good, heavy duty electric bug killer will do a super job on the adults. Work some Diazinon granules into your garden soil or use Oftanol on your lawn to get the grubs. Like I told you . . . trouble!

White Grub larva, pupa and adult
(VWR)

White Grubs in soil close up
(VWR)

White Grubs in soil and held in hand
(VWR)

Green June Beetle — larva
(USDA)

259

Green June Beetle
(DOF)

Pitch-eating Weevil Grub
(DOF)

_____Wireworms _____

Every gardener who has tilled the soil will have turned up some of these critters from time to time. Wireworms attack virtually every garden crop, damaging planted seed and plant roots. They will also bore into large roots, stems and tubers causing damage ranging from poor stands to complete loss.

Wireworms are larvae of Click Beetles. They come from a large family — over 800 species in North America. Wireworms, cylindrical and elongate in shape, are smooth, shiny and hard-bodied. Color varies from yellow to brownish. When you encounter one of these, kill it.

Note: If you want to know why Click Beetles are called click beetles, place one on a hard, smooth surface. Flip him over on his back and wait a minute.

Wireworms = Order: Coleoptera; Family: Elateridae

Wireworm in soil near sprouting corn seed
(USDA)

Wireworm or Click Beetle larva
(VWR)

Click Beetle adult (brown)
(VWR)

Click Beetle adult (black)
(VWR)

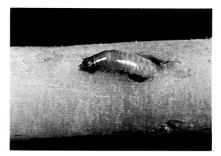

Wireworm. "Here she comes, Miss
America . . ."
(USDA)

Other Insect Critters

Critters that did not make the squad because of grades or disciplinary
reasons:

BRONZE LEAF BEETLE adult.
(DOF)

BRONZE LEAF BEETLE feeding
damage to eucalyptus leaves.
(DOF)

ORANGE DOG. Larva of Florida's giant swallowtail butterfly; it looks like something you might find under a chicken roost, but a few of these can strip a citrus tree.
(DOW)

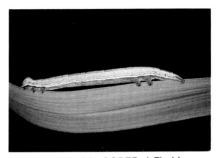

STRIPED GRASS LOOPER. A Florida lawn chewer.
(DOW)

ORANGE-HUMPED OAKWORM. This guy never knows which end is up!

Andrallus spinidens. No common name for this sucking insect which came from the Clewiston, Florida area. A good one to name-drop at parties so people will know you are smart.
(DOW)

VEDALIA BEETLE. This little Ladybird Beetle was imported from Australia to control Cottony Cushion Scale, and here it is doing that very thing on a Tung oil tree. Beneficial.
(DPI)

PARSLEYWORM. Larva of the Black Swallowtail Butterfly.
(DOW)

263

FOREST TENT CATERPILLAR. It looks like a big powder puff, but it really is a Forest Tent Caterpillar cocoon.
(VWR)

VELVET ANT. A parasite of wasps and bees.
(USDA)

DAMSEL BUGS "make it" off other insects; they are wholly predaceous.
(VWR)

KATYDIDS are grasshopper kinfolks; they mostly just make noise.
(VWR)

MAHOGANY WEBWORM or LEAF TIER occasionally causes extensive defoliation of host mahogany trees . . . unsightly but not fatal.
(DOF)

CLOVER MITE egg, young and adult.
(VWR)

264

CATALPA SPHINX. Fishermen cherish this caterpillar for bait. It's a big 'un, about 3 inches long. Big bait = big fish.
(SFI)

GRUBS are worse than little kids; here's one in a sweet potato.
(USDA)

SPINY OAKWORM is a southern species much like the native eastern Orange-striped Oakworm.
(DPI)

FRASS. This bug poop indicates upper-story damage.
(DOF)

SEVERE DEFOLIATION means this! Catalpa Worms did this number.
(DOF)

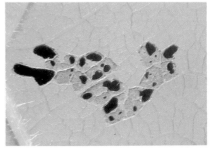

SKELETONIZING looks like this.
(DOF)

265

POPLAR TENT CATERPILLAR is also
known as the Poplar Tent Maker.
(DOF)

CATOCALA CATERPILLAR. Adults are
Underwings, an attractive group of
moths, specimens of which are prized
by all amateur collectors; check the
camouflage job on this one.
(DPI)

SYRPHID FLY adult . . . this picture is
too good not to use.
(VWR)

____Other Animal Critters____

As I reminded you earlier, less than one half of one percent of all insects are considered pests of plants; there are also two-legged and four-
legged pests of plants.

As few as four squirrels or four 'coons are capable of destroying and
disposing of 200 pounds of pecans. Mature trees, of course, can easily put

out 100 pounds or even much more a season, yet hungry predators such as raccoons, field rats and squirrels can easily wipe out the entire crop.

Two types of damage are caused by the ravages of squirrels . . . likely the FOX SQUIRREL or the GRAY SQUIRREL, for they are the more numerous; one is the removal or destruction of the nuts themselves, and this damage begins long before the nuts are mature and will continue until the nuts are all gone. The other type of damage performed by ol' "bushy-tail" is attacking and girdling young, tender shoots of trees. Budded or grafted limbs may be destroyed in the same manner. Vegetable garden produce may be devastated by rodents, and these critters like everything you like. TREE SQUIRRELS usually nest in trees, but are not above setting up housekeeping in the attic of your home. Their habits and damage are akin to those of rats.

Tree Squirrel
(VWR)

Tree Squirrel nest
(VWR)

Fox Squirrel almost hiding in small tree.
(Just washed his tail and can't do a
thing with it!)
(DOF)

Gray Squirrel feeding damage to
camphor tree
(DOF)

267

In the event you find a nut that is half gnawed, shell and nut-meat ... this will likely be squirrel damage; if you find a nut that is partially gnawed, but hollow ... this is likely the work of a FIELD RAT or FIELD MOUSE. Just after sundown, one evening, I shot what I thought was a gray squirrel that had lost all the hair on its tail. Know what it was? A big rat! This critter was scurrying along a limb of one of my pecan trees.

Field Mouse and runway
(ORTHO)

Many a cat has unjustly borne the wrath of the neighborhood for clawing big holes in loaded, plastic garbage bags that await the garbage truck, when all the while it was mischief performed by brother RAC-COON. These hardy little predators with their enormous appetites, ingenuity and dexterity in obtaining food are highly destructive to fruits and nuts prior to harvest. And, like the squirrels, they are attracted to buds, young fruit, tender shoots and blossoms. They can eat their weight many times over in produce in a single week. A family of these hungry little mammals can routinely destroy an entire nut harvest so everything is gone except an isolated nut or two.

Raccoon. 'Coons will eat your pecans,
but they are mostly scavengers and are
at their best in your garbage can.
(ORTHO)

DOGS are dynamite in the yard and garden. They will dig up freshly prepared flower beds to prepare a cool, soft place to sleep; they destroy turf by making trails where they repeatedly run; they might tear up a patch of lawn grass with their teeth and paws, digging to the sounds of soil insects; and they might even chew up ornamental and vegetable plants just for the hell-of-it. Male dogs instinctively mark their territory by "throwing up a leg" on chosen plants and scent-marking them with urine. This practice seldom hurts tree bark, but it sure will damage the leaves of many smaller plants and if repeated frequently, can kill the plant. (A polite way to discuss this problem is simply refer to it as K-9-P disease.)

Damage to Japanese boxwood hedge
by dog urine
(DOF)

BEAVERS can be wretches, not only with small saplings, but also with larger trees. They have a habit of severely damaging or killing random trees by girdling. These busy-bodies will clip seedling sprouts, sever stems of saplings close to the ground, and remove broad bands of bark and sapwood. If they drop a tree, they leave a gnawed, conical stump.

Beaver damage to pine tree
(DOF)

RABBITS seem to eat just about everything. Of course, you know an ol' cottontail can ruin more garden than he needs. Rabbits and hares may also injure or destroy ornamental shrubs, fruit trees, berry bushes and even less succulent pine seedlings and saplings. Br'er rabbit tends to leave an angled cut almost all the way through a seedling stem so that the top part flops over.

You know a Rabbit will graze your garden, but look at what one did to this little slash pine.
(DOF)

MOLES are small, furry mammals that live underground. Some species dig many shallow feeding runways only a few inches below ground, which produces ridges on the soil surface, in their quest for earthworms, slugs, grubs and other small insects. They seldom feed on plants; damage is caused by their tunneling which may uproot the plant or loosen the soil around roots to the extent they dry out and die. The best way to rid your lawn and garden of moles is to control the insects they feed on.

Mole
(ORTHO)

Mole ridges caused by tunneling
(ORTHO)

DEER do a lot of damage to a lot of plants. They feed on many vegetables and flowers; they may also browse on the tender bark, leaves and twigs of shrubs and trees. Bucks (males) have a habit of damaging trees by removing bark when rubbing velvet from their antlers. Some soybean farmers in Louisiana found marauding deer were repelled by using an egg spray; 1 to 1½ dozen eggs in 5 gallons of water per acre. Deer have a keen sense of smell and will shy away from the odor of decomposing eggs; the diluted egg-water mixture is, however, too weak for human noses to detect. A heavy rain will remove the protective coating and if you choose to try this, don't get egg shells in your sprayer.

Deer damage to pine sapling
caused by velvet-stripping
(DOF)

BIRDS eat a great many insects and are mostly beneficial, but they also cause a good deal of damage. Once they develop the habit of feeding in your garden, you will probably have to resort to netting or screening to exclude them. Many kinds of birds feed on seeds, fruit and berries. They scratch away at soft soil to unearth newly planted seed; they peck at seedlings and young tender leaves.

Mockingbirds will fight a tom-cat, but they will also decimate your tomato and strawberry crop.
(VWR)

Crow . . . found guilty of pecan theft
(VWR)

House Sparrows can wipe out a newly planted garden. They can be tough on not only shallow-planted seed, but on tender seed sprouts as well.
(VWR)

WOODPECKERS, with their jack-hammer-like bills, will severely scar the bark of a tree in their search for insect food and they are not above attacking the wood siding of your house. (Woodpeckers must get king-size headaches!)

Bark plates pecked off tree by Woodpecker feeding on ants
(DOF)

Typical Woodpecker nest in rotting tree
(VWR)

YELLOWSHAFTED FLICKERS are capable of leaving a patch of sap-oozing holes in a tree that, at first glance, might look like somebody "shot-gunned" it with a buck shot load.

Yellowshafted Flicker damage to a pine tree
(DOF)

YELLOW-BELLIED SAPSUCKERS leave a distinctive calling card — random lines of deep drill-holes in the tree bark. Many people are prone to blame borers for this damage. Not so!

Yellow-bellied Sapsucker damage. Notice the in-line holes.
(DOF)

Yellowshafted flickers and yellow-bellied sapsuckers are members of the Woodpecker family. Sapsuckers may peck into many trees before finding one with sap that has a high sugar content. Once the birds find a suitable tree, they are likely to visit it many times a day and feed on it year after year.

The LOGGERHEAD SHRIKE, also known as the BUTCHER BIRD, is very reminiscent of the Mocking Bird in both appearance and in flight behavior. Shrikes capture live prey . . . insects, mice, snakes, even other birds. When the victim is too large to swallow whole, the Shrike will impale it on a thorn or barb to assist him in tearing it up; he has a strong beak, but weak talons. When perched, this bird is best identified by its rakish black mask, short hooked beak, and black wings and tail. The top of the head and back are gray; the chest and belly are lighter.

American bird grasshopper impaled on barbed wire fence by a Loggerhead Shrike
(DOF)

GOPHER TORTOISES are unmistakable. The hind legs are round and blunt like an elephant's foot; the forelegs are flattened and are heavily scaled in front — ideal for digging. They spend much of the time in their burrows, but come out in the early morning or late evening to browse. They are basically vegetarian, although occasionally they will eat meat. Piles of excavated dirt that mark the entrances to their long tunnels are familiar sights when they are present. These "gophers" are found throughout most of Florida except the extreme southern Everglades. The flesh is delectable, and this tortoise is widely hunted for its meat.

Gopher Tortoise
(DOF)

In the states of Georgia and Florida, the POCKET GOPHER is commonly known as a SALAMANDER. They should never be confused with Moles. Florida Pocket Gophers are medium-sized rodents; they may vary in total length from 8 to 12 inches from nose to tip of tail. They generally have dark brown to blackish fur, but have a hairless tail. An outstanding characteristic is the fur-lined, external cheek pouches which are used for transporting root sections and other bits of food . . . and are never used for carrying soil to the ground surfaces as some people suspect. Pocket Gophers are basically vegetarians, but on occasion will eat snails, earthworms or beetles.

These animals have strong front limbs with long heavy claws which are used for digging. Individual burrow systems have been excavated to lengths well beyond 150 yards. The most easily identified damage is the presence of the surface mounds of excavated, loose soil. They can easily be classified as a "disaster" on a golf course . . . or your front lawn. Why are Pocket Gophers called Salamanders? Some early writings suggest the name salamander developed from a contraction of a local designation, "sandy mounder."

Pocket Gopher mound
(DOF)

Pine sapling leaning because of root damage by a Pocket Gopher
(DOF)

Close up of root gnawed by Pocket Gopher
(DOF)

COWS, HORSES and HOGS can do more garden damage than a hurricane. A good fence is the answer here, but don't leave the gate open.

Hog . . . "Who, me?"
(DOF)

Skunk. Ya' know, this critter would actually be a good ol' boy . . . if he just didn't smell so bad!
(ORTHO)

This Black Rat Snake in the tree won't hurt you . . . but it can cause you to hurt yourself.
(DOF)

Rattlesnake. Now, THIS one can hurt you! Run for the hoe . . . the one with the long handle.
(VWR)

MAN, WOMAN and CHILD (that's US, folks) are responsible for more damage to our planet's ecology than any other critter or critters. Lest we forget!

Beaver = Class: Mammalia; Order: Rodentia; Family: Castoridae
Deer = Class: Mammalia; Order: Artiodactyla; Family: Cervidae
Dog = Class: Mammalia; Order: Carnivora; Family: Canidae
Hog = Class: Mammalia; Order: Artiodactyla; Family: Suidae
Pocket Gopher = Class: Mammalia; Order: Rodentia; Family: Geomyidae
Gopher Tortoise = Class: Reptilia; Order: Testudines; Family: Testudinidae
Rabbit = Class: Mammalia; Order: Rodentia; Family: Leporidae
Squirrel = Class: Mammalia; Order: Rodentia; Family: Sciuridae
Pileated woodpecker, Yellowshafted flicker and Yellow-bellied sapsucker = Class: Aves; Order: Piciformes; Family: Picidae
Black rat snake = Class: Reptilis; Order: Squamata; Family: Colubridae

TREE SQUIRRELS: See RODENTS, page 182.

RACCOONS are mostly a nuisance because of their fondness for your garbage. Use a metal garbage can with a well secured lid. If you use plastic garbage bags, set them out in the morning on trash pick-up days; remember 'coons forage at night. Adequately screen your masonry fireplace chimney opening and keep the damper shut when not in use. These critters not only enjoy the shelter provided by an unused fireplace chimney, but a mama might even raise her family there. Even one raccoon inside your home can do unbelievable damage to your furnishings . . . and to you. And brother 'coon will hole up in your attic if he has access to it. Block ingress and egress points at late night when he's likely to be out marauding. A coon boxed up in your attic is BIG trouble. In addition to scattering your trash, raccoons will eat your garden produce and tree fruit. If these problems become serious, you might have to resort to trapping or shooting. Your local pest control company can provide you with this service.

DOGS cause the most intolerable problems when they belong to someone else. Let's face it . . . dogs will be dogs! If *your* animal is destructive, the choice is obvious . . . provide training or suffer the consequences. If your *neighbor's* dog is the culprit, speak with your neighbor about his responsibility.

BEAVERS are generally not a problem around a homestead, but if yours is the exception, contact your extension forester for information. He will be glad to point you in the right direction for control measures.

RABBIT control is best achieved by fencing with one-inch mesh chicken wire at least 30 inches high. Make sure the bottom is imbedded in the soil to prevent the rabbit from digging under it. Shrubs and trees may be individually protected with cylindrical guards fashioned from the same material.

DEER are capable of leaping an average height fence, so fencing might not be the best solution. Try the egg spray repellent. Those Louisiana farmers say it really works. Elimination by shooting is another option, but remember deer are protected by law, so check with your State Department of Fish and Game first. Warning! If you exercise this option, you might be disowned by little children for killing Bambi.

MOLES are best controlled by eliminating their food supply — grubs and worms. (See WHITE GRUBS, page 258.) Trapping or gassing are other options.

BIRDS are beneficial and should not be eliminated, but

rather deterred. Bird netting draped over a fruit tree, strawberry patch or tomato bush is usually quite adequate in your garden or orchard. When birds become a problem because of their feeding, numbers or nesting, control measures in a given situation usually are highly specialized and really shouldn't be generalized. Controlling problem birds is not a simple task. Don't hesitate to ask for help from a responsible source.

WOODPECKER problems are difficult to solve. Many exterminators will not even attempt control because woodpeckers are protected by law and solutions are not simple. Barriers are probably the best bet. Before you resort to elimination, check with your Extension Forester for help.

YELLOWSHAFTED FLICKERS and YELLOW-BELLIED SAPSUCKERS: Wrapping the damaged tree trunk with burlap is one solution to a tough control problem.

LOGGERHEAD SHRIKES: No control measures necessary . . . unless you are a frog or a snake.

GOPHER TORTOISE control? Eat it!

POCKET GOPHERS or Salamanders may be controlled by trapping or baiting or gassing. An active cat will often do the job. Many publications deal with the control of Pocket Gophers, so do some homework. Check with your local Extension Agent.

MAN, WOMAN and CHILD: Best control would hardly be *elimination*, but *education* would certainly be in order. And I would fondly hope this book aids that effort.

Appendix

CRITTER GROUPS

The following list is not intended to be an index of critters appearing in this book. True, most are shown in living color and are discussed in some detail, but as in the cast of a play, their frequency of appearance and importance to the plot will vary. Aphids are a good example. Aphids are like kittens . . . they can do mischief in many situations; borers, varieties notwithstanding, generally follow the same script if attacking a pecan tree or an oak tree. And critters are not necessarily villains; many are heroes and their presence should be encouraged. So, when you have a problem, perhaps this critter-cast will offer a clue as to where to look for the culprit.

BENEFICIAL CRITTERS

Preying Mantids
Assassin Bugs
Minute Pirate Bugs
Ladybird Beetles
Ground Beetles
A Soft-Winged Flower Beetle
Syrphid Flies
Lacewings
Ant Lion (Doodle Bug)
Bigeyed Bugs

Honeybee
Wild Bees
Parasitic Wasps
Predatory Wasps
Spiders
Bark Louse
Mealybug Destroyer
Predaceous Stink Bug
Tiger Beetle
Ambush Bugs

Sucking
Aphids
Leafhoppers
Scale Insects
Whiteflies
Spider Mites

Chewing
Thrips
Leaf and Flower Feeding Beetles:
Rose Chafers
Rose Leaf Beetles
Twelve-spotted Cucumber
Beetle
Rose Slugs
Leafrollers
Leafcutting Bees
Rose Stem Borers
Gall Wasps
Rose Midge
Grasshoppers
Leaf and Flower-Feeding Larvae

Citrus Rust Mites . . . wedge shaped body; lemon-yellow color. They can just barely be seen with the naked eye and are best seen with a 10-power magnifying glass.

Citrus Red Mites (Purple Mites) . . . bright red to deep purple and infest leaves, fruit and new growth. Most numerous from May through July, but can be most damaging from October through February.

Six-spotted Mites . . . live in colonies on the under surface of leaves and usually prefer grapefruit, but can be found on other varieties of citrus.

Texas Citrus Mites . . . brownish-green in color and are about the same size as citrus red mites.

Purple Scale and Glover Scale . . . are very similar in appearance and habits, but Glover scale is longer and narrower. They are often overlooked because they are found primarily on the inner parts of the tree.

Chaff Scale . . . forms a light brown nearly round armor closely resembling a piece of wheat chaff.

Citrus Snow Scale . . . is a serious pest in most of the citrus growing areas of Florida.

Florida Red Scale and Yellow Scale . . . are armored scales of similar size and shape.

Cottony Cushion Scale . . . the Vedalia ladybeetle feeds on this scale and usually keeps it under control.

Whiteflies . . . never spray when a large number of adults are present; wait 10 to 12 days until most have disappeared. This will allow enough time for the eggs to hatch and the young to be killed before they can cause much damage.

Citrus Blackflies . . . are close relatives of the common whitefly; if you find it on the underside of leaves, call your county agent.

Mealybugs . . . most common during spring and early summer.

Aphids . . . insecticides should be applied before the leaves curl.

Orange Dogs . . . may be quite destructive to young trees; pick by hand and step on them.

Grasshoppers and Katydids . . . will eat some leaves, but are no big threat.

COMMON HOUSE-INFESTING ANTS

Species	Nest Location	Preferred Food
Argentine	Lawns and plant beds	Sweets, animal fat
Crazy	Trash piles, rotten wood, soil, cavities in trees	Sweets, meat, grease, fruit
Fire	Lawns, plant beds, gardens	Meat, grease
Little Black	Lawns, under objects, rotten wood	Grease, sweets, meat, fruit
Pavement	Cracks in paving	Grease, meat, honey
Pyramid	Gardens and plant beds	Sweets
Acrobat	Wood and mortar	Meats, sweets
Carpenter	Stumps, logs, fences, homes	Sweets and nearly everything else
Thief	Nests of other ants, soil, cracks in walls	Meat, sweets, cheese, grease
Odorous	Under floors, stones, or in walls	Sweets, meats, dairy products
Pharaoh	Various locations near heat and moisture sources	Meats, grease, sweets

Orchids are big in Florida. The climate is for the most part favorable for growing these beauties; unfortunately, conditions which are favorable to orchids are also favorable to orchid pests. Here is a modified list of critters that you will likely contend with if your fancy is orchidology.

Scale Insects —
    Armored Scales:
        Boisduval scale
        Proteus scale
        Florida red scale
        Red orchid scale
        Vanda orchid scale
        Cactus scale
        Pineapple scale
        Tesserate scale
        Glover scale
        Orchid scale
        Black thread scale
        Latania scale
        Dictyospermum scale
        Coconut scale
        White peach scale
        Chaff scale

Soft Scales:
    Brown soft scale
    Stellate scale
    Orchid pit scale
    Hemispherical scale
    Black scale
    Acuminate scale
    Florida wax scale
    Pyriform scale

Mealybugs —
    Orchid mealybug
    Long-tailed mealybug
Flies
Leafhoppers
Planthoppers
Whiteflies
Bees
Grasshoppers
Crickets
Roaches
Lacewings
Earwigs
Millipedes
Springtails
True Bugs
Snails and Slugs
Katydids

Beetles —
    Dendrobium beetle
    Three-spotted lady beetle
Aphids —
    Melon aphid
    Palm aphid
Termites
Wasps —
    Orchidfly (Not a true fly)
Ants —
    Florida carpenter ant
    Little fire ant
    Thrips —
        Florida flower thrips
        Cuban-laurel thrips
        Greenhouse thrips
        Banded greenhouse thrips
        Red-banded thrips

Butterflies, Moths and
Skippers —
  The great southern white moth
  Io moth
  Morning sphinx
  Puss caterpillar
  Black swallowtail, parsleyworm
  A fruit scarring worm
  Southern armyworm
  Yellow-striped armyworm
  Pink scavenger caterpillar
  Saddleback caterpillar
  Cabbage looper
  Tortricid moths, larvae are leaf rollers

Mites —
  Phalaenopsis mite
  Tumid spider mite
  Two-spotted mite

Remember, this is only a partial list of critters that do bad things to orchids . . . give up? (Well, you can always switch to a nice patch of nut grass. Nothing seems to bother that.)

──────── PANTRY CRITTERS ────────────────────

Whole Grain Insects
  Angoumois Grain Moths
  Grain Weevils:
    Rice Weevils
    Granary Weevils
  Bean and Pea Weevils
  Cadelle Beetles

Miscellaneous Pests
  Brown Spider Beetle
  Mealworms
  Psocids (Booklice)

Processed and Dried-Food Insects
  Grain and Flour Beetles
  Cigarette and Drugstore Beetles:
    Confused Flour Beetles
    Saw-toothed Grain Beetles
    Cigarette Beetles
  Dermestid Beetles:
    Larder Beetles
    Carpet Beetles
    Cabinet Beetles
  Flour Moths:
    Indian Meal Moth
    Almond Moths
    Mediterranean Flour Moth

Attacking the nuts
Pecan Nut Casebearer
Hickory Shuckworm
Pecan Weevil
Stinkbugs and Plantbugs

Attacking the limbs, trunk and twigs
Pecan Twig Girdler
Red-Shouldered
   Shot-Hole Borer
Flatheaded Borers
Obscure Scale

Attacking the foliage
Aphids
Mites
Pecan Leaf Casebearer
Pecan Nut Casebearer
Pecan Cigar Casebearer
Pecan Phylloxera
Sawflies
May Beetles
Fall Webworm
Walnut Caterpillar
Pecan Catocala
Pecan Spittlebug
Pecan Bud Moth
Leafminers

PINE TREE CRITTERS

Bark Beetles
Southern Pine Beetles
Ips Engraver Beetles
Black Turpentine Beetles

Wood Borers
Southern Pine Sawyers
Turpentine Borers
Ambrosia Beetles

Root and Twig Insects

Bark Feeders
Pine Tip Moths
Pine Weevils
White Grubs

Leaf Feeders
Sawfly
Pine Webworm
Pine Colaspis Beetle
Pine Webworms

Miscellaneous Insects
Pine Needle Miner
Pine Pitch Midge
Scale Insects
Aphids
Spider Mites
Cypress Leaf Beetle
Southern Pine Coneworm

Conifers

Southern Pine Beetle
IPS Engraver Beetle
Black Turpentine Beetle
Pine Tip Moths
Reproduction Beetles
Wood Borers
Pine Sawflies
Pine Webworm
Bagworms
Pine Colaspis Beetle
Sawyer Beetles
Ambrosia Beetles
Termites
Scales
Southern Pine Coneworm
Deodar Weevil
Cypress Leaf Beetle

Hardwoods

Fall Webworms
Tent Caterpillars
Variable Oak Leaf Caterpillars
Elm Leaf Beetles
Walnut Caterpillars
Cankerworms
Twig Girdlers
Leaf Rollers and Leaf Tyers
Carpenter Worms
Carpenter Ants
Carpenter Bees
Walkingstick
Aphids
Wood Borers
Termites
Gall Insects
Scales
June Beetles
Catalpa Worm
Powder Post Beetles
Cottonwood Leaf Beetle
Gypsy Moth
Tussock Moth
Yellownecked Caterpillar
Clearwing Moths
Leafminers
Black Twig Borer
Treehoppers

VEGETABLE CRITTERS

Soil

Cutworms
Wireworms
White Grubs
Maggots

Rootworms
Sweet Potato Weevil
Mole Crickets
Pillsbugs and Sowbugs

Chewing

Corn Earworm
Cabbage Looper
Imported Cabbageworm
Diamondback Moth
Armyworms:
  Beet Armyworm
  Fall Armyworm
  Southern Armyworm
  Yellowstriped
    Armyworm
Tomato and Tobacco
  Hornworm
Tomato Pinworm
Leafminers
Squash Vine Borer

Pickleworm
Saltmarsh Caterpillar
Garden Webworm
Beet Webworm
Cowpea Curculio
Vegetable Weevil
Colorado Potato Beetle
Tortoise Beetles
Flea Beetle
Mexican Bean Beetle
Cucumber Beetle
Squash Beetle
Blister Beetle
Grasshoppers
Melonworm

Sucking

Squash Bug
Harlequin Bug
Stink Bugs
Leaffooted Bugs
Garden Fleahopper

Leafhoppers
Aphids
Thrips
Mites
Whitefly

———— WOOD-DESTROYING CRITTERS ————

Powderpost Beetles
Death Watch Beetles
Bark Beetles
Flatheaded Borers
Old House Borers
Flat Oak Borer
Other Roundheaded Borers

Drywood Termites
Subterranean Termites
Timber Worms
Carpenter Ants
Carpenter Bees
Spider Beetles
Weevils, Snout Beetles

___Index_____

A

Abbot's sawflies, 191-193
ambrosia beetles, 28-29
American cockroaches, 88-93
American dog ticks, 226-227
angoumois grain moths, 167-172
animal critters, 266-276
antlions, 15-16
ants
    fire ants, 6-9
    common house-infesting, 280
    Florida carpenter ants, 5-6
    Florida harvester ants, 9-10
    household ants, 10-12
    little fire ant, 12-13
    pharaoh ants, 14
aphid-induced galls, 120
aphids, 16-18
aphids, rose, 187
aphytis wasps, 233
Apopka weevils, 243-244
armyworms, 19-20
assassin bugs, 20-21
Australian cockroaches, 88-93

B

bagworms, 21-23
baldfaced hornets, 234-235
bark louse, psocids, 23-24
beavers, 269, 277
bees
    bumble bees, 26-27
    carpenter bees, 26-27
    honeybees, 25

    leafcutting bees, 26-27, 188
    wild bees, 26-27
beetle-induced galls, 120
beetles
    ambrosia, 28-29
    black turpentine, 42-45
    blister, 30-31
    bronze leaf, 262
    brown spider, 168-172
    cabinet, 167-172
    carpet, 167-172
    cigarette, 167-172
    click, 260-262
    Colorado potato, 31-32
    cottonwood leaf, 32-33
    cucumber, 33-34
    cypress leaf, 35-36
    deathwatch, 52-54
    dermestid, 167-172
    drugstore, 167-172
    elm leaf, 36-37
    flatheaded, 60-62
    flea, 37-38
    flour, 167-172
    furniture, 52-54
    grain, 167-172
    ground, 38-39
    Ips engraver, 42-45
    ladybird, 40
    larder, 167-172
    Mexican bean, 41
    old house borer, 52-54
    pine bark, 42-45
    pine colaspis, 46-47
    powderpost, 52-54
    rose leaf, 188

NOTES

I know, ''Don't write in the book,'' we have been taught;
but do it . . . and do it here!

NOTES

NOTES